Building Hawaii's Innovation Economy

Summary of a Symposium

Charles W. Wessner, *Rapporteur*

Committee on Competing in the 21st Century:
Best Practice in State and Regional Innovation Initiatives

Board on Science, Technology, and Economic Policy

Policy and Global Affairs

NATIONAL RESEARCH COUNCIL
OF THE NATIONAL ACADEMIES

THE NATIONAL ACADEMIES PRESS
Washington, D.C.
www.nap.edu

THE NATIONAL ACADEMIES PRESS   500 Fifth Street, NW   Washington, DC 20001

NOTICE: The project that is the subject of this report was approved by the Governing Board of the National Research Council, whose members are drawn from the councils of the National Academy of Sciences, the National Academy of Engineering, and the Institute of Medicine. The members of the committee responsible for the report were chosen for their special competences and with regard for appropriate balance.

This study was supported by Contract/Grant No. DE-DT0000236, TO# 28, (base award DE-AM01-04PI45013), between the National Academy of Sciences and the Department of Energy; and Contract/Grant No. N01-OD-4-2139, TO# 250, between the National Academy of Sciences and the National Institutes of Health. This report was prepared by the National Academy of Sciences under award number SB134106Z0011, TO# 4 (68059), from the U.S. Department of Commerce, National Institute of Standards and Technology (NIST). This report was prepared by the National Academy of Sciences under award number 99-06-07543-02 from the Economic Development Administration, U.S. Department of Commerce. The statements, findings, conclusions, and recommendations are those of the author(s) and do not necessarily reflect the views of the National Institute of Standards and Technology, the Economic Development Administration, or the U.S. Department of Commerce. Additional support was provided by the Heinz Endowments, the Association of University Research Parks, Acciona Energy, Dow Corning, IBM, and SkyFuel, Inc.

Any opinions, findings, conclusions, or recommendations expressed in this publication are those of the author(s) and do not necessarily reflect the views of the organizations or agencies that provided support for the project.

International Standard Book Number 13: 978-0-309-25663-6
International Standard Book Number 10: 0-309-25663-1

Limited copies are available from Board on Science, Technology, and Economic Policy, National Research Council, 500 Fifth Street, NW, W547, Washington, DC 20001; 202-334-2200.

Additional copies of this report are available from the National Academies Press, 500 Fifth Street, NW, Keck 360, Washington, DC 20001; (800) 624-6242 or (202) 334-3313; *http://www.nap.edu.*

THE NATIONAL ACADEMIES
Advisers to the Nation on Science, Engineering, and Medicine

The **National Academy of Sciences** is a private, nonprofit, self-perpetuating society of distinguished scholars engaged in scientific and engineering research, dedicated to the furtherance of science and technology and to their use for the general welfare. Upon the authority of the charter granted to it by the Congress in 1863, the Academy has a mandate that requires it to advise the federal government on scientific and technical matters. Dr. Ralph J. Cicerone is president of the National Academy of Sciences.

The **National Academy of Engineering** was established in 1964, under the charter of the National Academy of Sciences, as a parallel organization of outstanding engineers. It is autonomous in its administration and in the selection of its members, sharing with the National Academy of Sciences the responsibility for advising the federal government. The National Academy of Engineering also sponsors engineering programs aimed at meeting national needs, encourages education and research, and recognizes the superior achievements of engineers. Dr. Charles M. Vest is president of the National Academy of Engineering.

The **Institute of Medicine** was established in 1970 by the National Academy of Sciences to secure the services of eminent members of appropriate professions in the examination of policy matters pertaining to the health of the public. The Institute acts under the responsibility given to the National Academy of Sciences by its congressional charter to be an adviser to the federal government and, upon its own initiative, to identify issues of medical care, research, and education. Dr. Harvey V. Fineberg is president of the Institute of Medicine.

The **National Research Council** was organized by the National Academy of Sciences in 1916 to associate the broad community of science and technology with the Academy's purposes of furthering knowledge and advising the federal government. Functioning in accordance with general policies determined by the Academy, the Council has become the principal operating agency of both the National Academy of Sciences and the National Academy of Engineering in providing services to the government, the public, and the scientific and engineering communities. The Council is administered jointly by both Academies and the Institute of Medicine. Dr. Ralph J. Cicerone and Dr. Charles M. Vest are chair and vice chair, respectively, of the National Research Council.

www.national-academies.org

Committee on
Competing in the 21st Century:
Best Practice in State and Regional Innovation Initiatives[*]

_____

[*]As of January 2011.

PROJECT STAFF

Charles W. Wessner
Study Director

Sujai J. Shivakumar
Senior Program Officer

Alan Anderson
Consultant

David S. Dawson
Senior Program Assistant

McAlister Clabaugh
Program Officer

David E. Dierksheide
Program Officer

For the National Research Council (NRC), this project was overseen by the Board on Science, Technology, and Economic Policy (STEP), a standing board of the NRC established by the National Academies of Sciences and Engineering and the Institute of Medicine in 1991. The mandate of the STEP Board is to advise federal, state, and local governments and inform the public about economic and related public policies to promote the creation, diffusion, and application of new scientific and technical knowledge to enhance the productivity and competitiveness of the U.S. economy and foster economic prosperity for all Americans. The STEP board and its committees marshal research and the expertise of scholars, industrial managers, investors, and former public officials in a wide range of policy areas that affect the speed and direction of scientific and technological change and their contributions to the growth of the U.S. and global economies. Results are communicated through reports, conferences, workshops, briefings and electronic media subject to the procedures of the National Academies to ensure their authoritativeness, independence, and objectivity. The members of the STEP Board* and the NRC staff are listed below:

---

*As of January 2011.

STEP STAFF

Contents

Preface

Responding to the challenges of fostering regional growth and employment in an increasingly competitive global economy, many U.S. states and regions have developed programs to attract and grow companies as well as attract the talent and resources necessary to develop innovation clusters. These state and regionally based initiatives have a broad range of goals and increasingly include significant resources, often with a sectoral focus and often in partnership with foundations and universities. These are being joined by recent initiatives to coordinate and concentrate investments from a variety of federal agencies that provide significant resources to develop regional centers of innovation, business incubators, and other strategies to encourage entrepreneurship and high-tech development.

PROJECT STATEMENT OF TASK

An ad hoc committee, under the auspices of the Board on Science, Technology, and Economic Policy (STEP), is conducting a study of selected state and regional programs in order to identify best practices with regard to their goals, structures, instruments, modes of operation, synergies across private and public programs, funding mechanisms and levels, and evaluation efforts. The committee is reviewing selected state and regional efforts to capitalize on federal and state investments in areas of critical national needs. This review includes both efforts to strengthen existing industries as well as specific new technology focus areas such as nanotechnology, stem cells, and energy in order to better understand program goals, challenges, and accomplishments.

As a part of this review, the committee is convening a series of public workshops and symposia involving responsible local, state, and federal officials and other stakeholders. These meetings and symposia will enable an exchange of views, information, experience, and analysis to identify best practice in the range of programs and incentives adopted.

Drawing from discussions at these symposia, fact-finding meetings, and commissioned analyses of existing state and regional programs and technology focus areas, the committee will subsequently produce a final report with findings and recommendations focused on lessons, issues, and opportunities for complementary U.S. policies created by these state and regional initiatives.

THE CONTEXT OF THIS PROJECT

Since 1991, the National Research Council, under the auspices of the Board on Science, Technology, and Economic Policy, has undertaken a program of activities to improve policymakers' understandings of the interconnections of science, technology, and economic policy and their importance for the American economy and its international competitive position. The Board's activities have corresponded with increased policy recognition of the importance of knowledge and technology to economic growth.

One important element of STEP's analysis concerns the growth and impact of foreign technology programs.[1] U.S. competitors have launched substantial programs to support new technologies, small firm development, and consortia among large and small firms to strengthen national and regional positions in strategic sectors. Some governments overseas have chosen to provide public support to innovation to overcome the market imperfections apparent in their national innovation systems.[2] They believe that the rising costs and risks associated with new potentially high-payoff technologies, and the growing global dispersal of technical expertise, underscore the need for national R&D programs to support new and existing high-technology firms within their borders.

Similarly, many state and local governments and regional entities in the United States are undertaking a variety of initiatives to enhance local economic development and employment through investment programs designed to attract knowledge-based industries and grow innovation clusters.[3] These state and regional programs and associated policy measures are of great interest for their potential contributions to growth and U.S. competitiveness and for the "best practice" lessons they offer for other state and regional programs.

---

[1]National Research Council, *Innovation Policies for the 21st Century: Report of a Symposium,* Charles W. Wessner, ed., Washington, DC: The National Academies Press, 2007.

[2]For example, a number of countries are investing significant funds in the development of research parks. For a review of selected national efforts, see National Research Council, *Understanding Research, Science and Technology Parks: Global Best Practices: Report of a Symposium,* Charles W. Wessner, ed., Washington, DC: The National Academies Press, 2009.

[3]For a scoreboard of state efforts, see Robert Atkinson and Scott Andes, *The 2010 State New Economy Index: Benchmarking Economic Transformation in the States,* Kauffman Foundation and ITIF, November 2010.

STEP's project on State and Regional Innovation Initiatives is intended to generate a better understanding of the challenges associated with the transition of research into products, the practices associated with successful state and regional programs, and their interaction with federal programs and private initiatives. The study seeks to achieve this goal through a series of complementary assessments of state, regional, and federal initiatives; analyses of specific industries and technologies from the perspective of crafting supportive public policy at all three levels; and outreach to multiple stakeholders. The overall goal is to improve the operation of state and regional programs and, collectively, enhance their impact.

THIS SUMMARY

The symposium reported in this volume convened state officials and staff, business leaders, and leading national figures in early-stage finance, technology, engineering, education, and state and federal policies to review challenges, plans, and opportunities for innovation-led growth in Hawaii. The symposium included an assessment of Hawaii's natural, industrial, and human resources; identification of key sectors and issues; and a discussion of how the state might leverage national programs to support its economic development goals.

This summary includes an introduction that highlights key issues raised at the meeting and a summary of the meeting's presentations. This workshop summary has been prepared by the workshop rapporteur as a factual summary of what occurred at the workshop. The planning committee's role was limited to planning and convening the workshop. The statements made are those of the rapporteur or individual workshop participants and do not necessarily represent the views of all workshop participants, the planning committee, or the National Academies.

ACKNOWLEDGMENTS

On behalf of the National Academies, we express our appreciation and recognition for the insights, experiences, and perspectives made available by the participants in this meeting. We are especially indebted to M.R.C. Greenwood, President of the University of Hawaii, for her leadership in organizing the event, identifying topics, and generating interest across a broad spectrum of participants. We are also grateful to Alan Anderson for preparing the draft introduction and summarizing the proceedings of the meeting and to Sujai Shivakumar and David Dierksheide of the STEP staff for preparing the report manuscript for publication.

SPONSORS

We are grateful to our project and event sponsors and extend particular recognition to the National Institute of Standards and Technology, the Department of Energy, the Economic Development Administration, the National Cancer

Institute, the Heinz Endowments, the Association of University Research Parks, Acciona Energy, Dow Corning, IBM, and SkyFuel, Inc. for their support of the program. For the Hawaii conference, special recognition goes to the University of Hawaii Foundation, Hawaiian Electric Company, the Weinman Innovation Fund, American Savings Bank, the Queen's Medical Center, HiBEAM (Hawaii Business and Entrepreneur Acceleration Mentors), the Intellisis Corporation, HDTC (High Technology Development Corporation), the Hawaii Business Roundtable, Inc., Rainbowtique of the University of Hawaii, and Gebco Hawaii.

NATIONAL RESEARCH COUNCIL REVIEW

This report has been reviewed in draft form by individuals chosen for their diverse perspectives and technical expertise, in accordance with procedures approved by the National Academies' Report Review Committee. The purpose of this independent review is to provide candid and critical comments that will assist the institution in making its published report as sound as possible and to ensure that the report meets institutional standards for quality and objectivity. The review comments and draft manuscript remain confidential to protect the integrity of the process.

We wish to thank the following individuals for their review of this report: Saul Behar, Philadelphia Science Center; Keiki-Pua Dancil, Bio-Logical Capital; Harold Masumoto, Pacific International Center for High Technology Research; Stephanie Shipp, Institute for Defense Analyses; and Donna Vuchinich, University of Hawaii Foundation.

Although the reviewers listed above have provided many constructive comments and suggestions, they were not asked to endorse the content of the report, nor did they see the final draft before its release. Responsibility for the final content of this report rests entirely with the rapporteur and the institution.

Charles W. Wessner   Mary L. Good

I

OVERVIEW

Hawaii's Innovation Strategy

The belief that science-based innovation fuels job creation, economic growth, and competitiveness is encouraging policymakers in nations and regions around the world to develop their own innovation strategies. These strategies seek to leverage the skills and knowledge found in research universities to create and attract companies, train and retain the talent needed to support a knowledge-based economy, and develop new capabilities in areas such as renewable energy and health care. Furthermore, these strategies employ programs that are built on partnerships among universities, private firms, government research facilities, foundations, and community organizations.

Recognizing the opportunities for growth based on innovation, the state of Hawaii is actively seeking to diversify its economy by drawing on the University of Hawaii system and other research and educational organizations as engines of sustainable, innovation-led growth. To this end, the University of Hawaii (UH) under the leadership of its new President, Dr. M.R.C. Greenwood, convened an Innovation Council made up of nationally recognized experts to develop recommendations to grow the state's knowledge-based economy. In launching the Council's report, the University of Hawaii partnered with the National Academies Board on Science, Technology, and Economic Policy to convene a major conference in January 2011. This conference drew together the state's political, academic, and business leadership, along with federal officials and national experts to highlight key challenges and opportunities facing Hawaii and to identify key steps toward meetings the Innovation Council's objectives. This volume is a report of that conference, and this chapter provides an overview of key issues raised over the course of this two-day conference.

> "We have what it takes. We have the intellect, the inquisitiveness, and an entrepreneurial know-how to invent and incubate a knowledge sector in Hawaii."
>
> "Over these many years, I've been honored to support the University of Hawaii's research endeavors and to encourage the growth of our technology sector. It is my hope that this conference will encourage a greater intersection and connection between the two."
>
> Senator Daniel Inouye
> Keynote address

DIVERSIFYING HAWAII'S ECONOMY

Hawaii's unique location in the middle of the Pacific Ocean provides unique challenges as well as important opportunities. On one hand, the Hawaiian Islands are remote from the U.S. mainland, small geographically, with a population of nearly a million people. On the other hand, as Senators Inouye and Akaka pointed out in their conference keynotes, the islands are strategically located as America's "front door" to the vibrant economies of East Asia and are home to unique geographical features and land and marine life, as well as a rich cultural heritage.

Indeed, the participation of U.S. Senators Inouye and Akaka, as well as Governor Abercrombie, Lieutenant Governor Schatz, and U.S. Representatives Hirono and Hanabusa at the conference telegraphed the strong commitment of Hawaii's leadership to develop the state's economy through investments in knowledge-based growth.

The need to diversify and grow Hawaii's economy was underscored by Dr. Carl Bonham of the University of Hawaii Economy Research Organization (UHERO), who reviewed the islands' history of economic dependence on a series of single-products. This, he said, began some two centuries ago with the domination of the sandalwood market and continued through export economies based successively on whaling, sugar, pineapple, the military, and, currently, tourism. The state is, and will probably continue to be, a popular destination for visitors worldwide who seek a reliably balmy climate in an exotic setting, but the growth potential for traditional tourism is limited by transportation, lodging, and other constraints. Dr. Bonham also noted that the island has also relied on the support of military bases since World War II, but the contribution of this sector is unlikely to grow significantly.

While the state will certainly continue to benefit from visitor-based revenues and military spending, the challenge for this century, he said, is to develop the multi-sector, interdisciplinary capacities needed to generate S&T-based innovation and diversify the state's opportunities for economic growth.

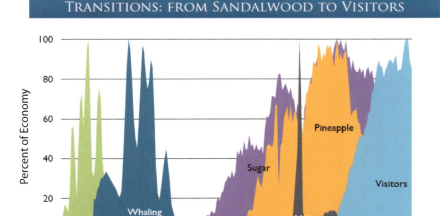

FIGURE 1 Hawaii's predominant industries, 1806-2009.
SOURCE: Carl Bonham, Presentation at January 13-14, 2011, National Academies Symposium on "E Kamakani Noi'i (Wind that seeks knowledge)."

THE UNIVERSITY OF HAWAII AS AN ENGINE OF GROWTH

In her conference presentation, University of Hawaii President M.R.C. Greenwood noted that her institution is seeking to become a major driver of innovation and commercialization in the state by initiating an institution-wide effort to enhance awareness of the potential applications of research and encourage entrepreneurship in its curriculum design, course content, and faculty incentives.

Setting Objectives

Dr. Greenwood listed three objectives in defining her institution's mission. The first objective, she said is to increase the number of educated citizens through the Hawaii Graduation Initiative. The second is to "create a 21st-century capability for innovation and technology transfer" supported by "a billion-dollar research industry for Hawaii." The final objective is to renovate and rebuild the university's infrastructure and to stimulate the creation of small businesses based on University of Hawaii technology developments.

Drawing on Expert Advice

In seeking to fulfill this entrepreneurial vision, Dr. Greenwood sought expert advice from both within and outside the state. In April 2010, President Greenwood formally named a University of Hawaii Innovation Council and charged its members with recommending how the university could best catalyze its own transformation and become a true leader in building regional economic leadership. The Innovation Council members included University of Hawaii faculty leaders, venture capitalists, a former Administrator of NASA, a dean at the University of California at San Diego, and the director of Stanford University's technology licensing office. The Council responded with a bold set of recommendations,[1] advocating that the university should:

- Identify research as an *industry* in Hawaii;
- Establish a Hawaii Innovation Technology Exchange Institute (HiTEx);
- Identify areas for commercialization opportunities; and
- Integrate entrepreneurship into the curriculum.

Partnering with Government

A key element of the Innovation Council's advice was to strengthen the university's partnerships with other sectors, including the state's political leadership. For example, Dr. Greenwood said that the University and Hawaii Governor Neil Abercrombie shared many views on how best to strengthen the state's innovation capabilities.

In his luncheon remarks to the conference, the first major speech on economic development since his inauguration a few weeks earlier, Governor Abercrombie's emphasized the need for a strong partnership between the University of Hawaii and the state government. He noted that his own campaign document, "A New Day in Hawaii," envisioned a key role in development for the university, explicitly encouraging greater involvement for university faculty: "Support the entrepreneurial professor, facilitate innovation and technology transfer, support premier education and research projects."[2]

Incentivizing Faculty

Several symposium participants also drew attention to the desirability of providing more incentives for entrepreneurial and outreach activity by faculty members, emphasizing that entrepreneurship is found not only in technology, but

---

[1] Access the Final Recommendations of the President's Innovation Council Report at <*http://www. hawaii.edu/offices/op/innovation/council-final-recommendations.pdf*>.

[2] Access the Campaign Document "New Day in Hawaii" at <*http://newdayhawaii.org/pdf/ ANewDayinHawaii.pdf*>.

also in social networking, the arts, social sciences, and virtually all professions. In this regard, the dean of the University of Hawaii Medical School, Dr. Jerris Hedges, said that the process of promotion and tenure at his institution had been reviewed, leading to a decision to shift the reward system toward teamwork and partnerships of various kinds. "We want to reinforce that the landscape has changed and we're expecting a different level of performance on the part of our faculty." A related incentive, he said, was to reward contributions to collaborative grants and accomplishments in the private sector, such as the generation of patents and work with start-up companies.

Partnering with Local Communities

Such partnerships between large firms, small firms, universities, and government can yield tangible results. In his presentation, Dr. Luis Proenza, President of the University of Akron described the steps taken at his university not only to draw out and capitalize on existing resources and build new capacities, but also to reach out to the community and to create innovative structures that strengthen both educational and economic activities. For example, he noted that the University of Akron had forged partnerships with the city's hospitals around broad issues, such as community renovation and creation of a privately run university foundation that supports licensing. "We are no longer alone as a university," he said, "but a key partner in the community and its knowledge economy."

BUILDING HAWAII'S INNOVATION ECONOMY

Recognizing Hawaii's unique assets as well as challenges, conference participants identified a series of opportunities and challenges in developing Hawaii's potential for knowledge-based growth.

Securing Venture Capital

Moving ideas from the laboratory to the marketplace requires more efficient flows of capital to young firms, and these flows were not yet adequate in Hawaii, said Dr. Wuh, an entrepreneur and a member of the University of Hawaii Innovation Council. To succeed, he said, entrepreneurs have to "follow the money. What we don't want to do is create a society of great entrepreneurs only to see them leave the state because there are insufficient resources to support their businesses."

In his conference presentation, Mr. Weinman, a venture capitalist and a fellow member of the University of Hawaii Innovation Council, pointed out that some of the largest pools of private capital in Hawaii, including those of the Kamehameha Schools and the labor unions, are invested not in the state itself but on the mainland. One outcome of this custom, he said, is to deprive young firms of financial opportunities locally. The need for private engagement is especially great, said Mr. Weinman, in an era when the budgets of public universities are declining. "We see an urgent need for a seed capital fund for the university," he said, "to put money into researchers with an entrepreneurial bent." If this fund is privatized, he added, it would be able to act with the speed and initiative of a private company. While quick decisionmaking is seldom appropriate or even possible for universities, it is a necessity for small firms. "If you want the university to be a research business," he said, "you need to give it some of the freedoms of a research business."

Incentivizing Risk Taking

A reluctance to take risks can inhibit a culture of entrepreneurship. But as some conference participants pointed out, this aversion to risk is often rooted in and reinforced by outmoded institutional practices. In his conference presentation, Dr. Wuh noted that the acceptance of risk, although foreign to the community of university researchers, is well accepted and even necessary in a modern entrepreneurial culture. As he noted, "Failure is what you learn from."

In his remarks, Mr. Weinman added, "The University of Hawaii must move beyond the old ways of doing things, regardless of how well they fit the old mold of academia. These ways are too slow for entrepreneurs. We need to have people working at the University of Hawaii to go beyond research and help create spillovers and entrepreneurial activity. We have to change the thinking. In Silicon Valley, everyone wants to be an entrepreneur." Another speaker, Daniel Golden, agreed with this prescription, adding, "The University of Hawaii's task is not [just] to be the center of this [innovation] system, but to help build it."

Attracting Star Talent

In his presentation, Art Ushijima highlighted the initiative taken by Hawaii's Queen's Medical Center to reach out to the mainland and Europe to attract several established "star" researchers to its staff and to help them establish facilities and infrastructure. One advantage of recruiting senior researchers, he said, is that they bring with them not only expertise, but also grants, graduate students, and collaborators. They are also likely to bring a degree of "path to market" expertise, a current focus of Queen's, which is seeking to raise its own return on investment.[3]

---

[3]This is the path taken by Singapore in its efforts to recruit star faculty and Nobel laureates. It is also the approach of the Canada Research Chairs program, which allocates $300 million a year to recruit internationally recognized leaders in their disciplines as well as exceptional emerging researchers seen by their peers to have potential to lead in their fields. Access the Government of Canada Web site at <*http://www.chairs-chaires.gc.ca/*>.

In doing so, Mr. Ushijima said, the hospital collaborates with the venture community and the medical school, seeking expertise as needed and building on the experiences of others. This initiative, he said, is already paying off in the form of new uses for established drugs, a novel solution to the blurring that often impairs MRI imaging, and a new use of PET scanning to visualize prostate cancer. The hospital has also developed a new approach to supporting research by setting aside a portion of indirect funds to generate interest until they are needed.

Encouraging Civic Involvement

In her conference presentation, Mary Walshok described how the community in San Diego, California, successfully mobilized to transform a region with an economy dependant on the presence of the military into one of the world's most productive clusters of science-based economic activity. The rapid growth of research and research-based firms began around a core of outstanding research institutions: Scripps Institution of Oceanography (1903), General Atomics (1955), University of California at San Diego (UCSD, 1960), and the Salk Institute (1960). Today this core has grown to more than 50 research institutes, including the Sanford-Burnham Medical Research Institute, with substantial research budgets.

Dr. Walshok said that San Diego provides a model for Hawaii in many respects. San Diego, she noted, began its innovation efforts in the 1960s when civic leaders wisely zoned land near the city's major research institutions to favor young innovative companies. In turn, the city's major institutions—including Salk, UCSD, and Scripps—used a strategy of hiring star researchers to develop a network of talented scientists in the region. In addition, universities developed curricula to develop the skills needed by emerging technology companies in the region. She cited the example of Irwin Jacobs, founder of Qualcomm, who needed engineers trained in his new wireless technology, CDMA, which was not being taught at other engineering schools. UCSD complied, and the company succeeded.[4] To overcome a relative paucity of investment capital, the city's citizens themselves raised small amounts of seed money to attract the first new firms. Most of all, said Dr. Walshok, the region capitalized on the strong commitment to place of those who lived there to generate enormous volunteer energy.

---

[4]Qualcomm illustrates the value of leveraging federal programs such as the Small Business Innovation Research Progam (SBIR). At a critical moment in the development of CDMA technology, Qualcomm received a series of SBIR awards that, according to the Qualcomm founder Irwin Jacobs, gave the firm credibility with investors and enabled them to further develop in this promising but at the time unproven technology. Qualcomm chips are, of course, the backbone for wireless communication, and the company now has nearly 18,000 employees, a market value of $90 billion, and has generated some 26,000 related jobs in the San Diego region. See Testimony by Irwin Jacobs before the Senate Small Business and Entrepreneurship Committee on February 19, 2011. In addition, see below the reference to SBIR by White House National Economic Council Advisor, Ginger Lew.

In his presentation, Robert McLaren of the University of Hawaii Institute of Astronomy noted that the University of Hawaii's Mauna Kea observatories have built strong links to the community.[5] While the University's Institute of Astronomy determines the direction of the scientific program, as it has done since its formation in 1967, all other aspects of governance are now managed by the Mauna Kea Management Board, a community group that addresses environmental, cultural, and access issues. The planned Thirty-Meter Telescope (TMT), which will have nine times the light-collecting area of today's largest telescopes, will be the first astronomical project to undergo the Management Board's complete project review process, from the initial planning stages through construction and operation.

Facilitating Technology Transfer from Universities

Discussing how to move technologies into useful applications, Katherine Ku of Stanford's technology licensing office told the symposium that despite Stanford's success, the process of technology transfer is inherently slow and unpredictable, and did not respond well to a strategy of "chasing the dollars." The higher good, she said, was to make new ideas useful. "We don't want the technologies to lie fallow in the literature. I think we need the inventors to be championing the technology, and we want to help them. Our goal is not just to make money."

Another difficulty experienced by those who work in technology transfer offices, especially in state universities, was the feeling that "we can't afford to fail." "There are too many restrictions on those individuals," said Barry Weinman. "These offices frustrate the innovators to the point where they don't want to deal with them." He firmly endorsed the use of a "501(c)3 strategy" to remove technology transfer in public universities, such as the UH, from the oversight of the state. Under an independent entity, he said, each technology can be addressed wholly on the basis of its commercial merits.

Building on Natural Assets

Several conference participants noted that when moving ahead with innovative development strategies, citizens of Hawaii are well aware of the potential of their state's abundant natural assets, such as sunlight and wind, clean ocean water, geothermal energy sources, and its mountain peaks that serve as home to some of the world's premier astronomical sites.

---

[5]When the first major observatory was constructed on Mauna Kea in 1970, the University of Hawaii did not yet have a night-time astronomy program (it had one solar observatory on Haleakala, Maui). It nonetheless edged out both Harvard and the University of Arizona in winning NASA awards to construct a pioneering 2.2-meter observatory. A key step in this success was creation of the Institute of Astronomy, which continues to oversee the scientific programs of the 13 observatories, today the largest such complex in the world.

In his presentation, Mr. Maurice Kaya of the Hawaii Renewable Energy Development Venture described the Hawaii Clean Energy Initiative, begun in 2009, as an effort to both conserve available resources and develop new, indigenous sources from wind, sun, geothermal heat, and biomass.[6] He said that this plan has the objective of reducing the consumption of fossil fuels in Hawaii, which now account for 90 percent of energy consumption. The state intends to reduce fossil fuel consumption by 70 percent by the year 2030—an "unheard-of" goal, he said. Describing this goal as "economically essential," Richard Rosenbloom, President of Hawaiian Electric Company noted that of a Gross State Product of $60 billion, more than $8 billion is spent on fossil fuels—a level deemed unsustainable by the state.

According to Dr. Taylor, the advantages of developing Hawaii's natural energy resources reach far beyond energy savings and pollution reduction. By developing new technologies in Hawaii rather than importing them from elsewhere, he said, the people of Hawaii would benefit locally by generating high-paying jobs, new careers for youth, energy security, and opportunities to export innovative technology, while diversifying income streams for owners of agricultural land. "We don't know what the best technologies will prove to be," said Dr. Taylor, "but when we do, we can teach others, and export knowledge instead of pineapples."

Leveraging Existing Assets

While the Hawaiian islands continue to appeal to tourists around the world, this industry's growth potential is limited, said Peter Ho, Chairman, President, and CEO of the Bank of Hawaii, and Chairman of the APEC 2011 Hawaii Host Committee. To maximize on its existing hotel stock, the state is seeking to draw more diverse business travelers at times of year when traditional tourism slows. Speaking at the roundtable session of the conference, Mr. Ho encouraged greater efforts to market the state for business and professional conventions, including from the economies of the Pacific Rim. Moreover, from this increased flow of business leaders and skilled people, Mr. Ho predicted that the state could expect new kinds of business collaborations to grow from APEC partnerships.

According to Brian Taylor of the University of Hawaii School of Ocean and Earth Science and Technology, another long-existing resource that can be further developed is the Pacific Missile Range Facility (PMRF) on Kauai. To date, this site has been used primarily by NASA, off-island companies, and other nations to launch satellites into Earth orbit. However, the Hawaii Space Flight Laboratory

---

[6]To prepare for this initiative, the state has also issued *Hawaii's Green Workforce: A Baseline Assessment* (Department of Labor and Industrial Relations, Research and Statistics Office, December 2010). This survey estimates the number of jobs that contribute to environmental protection, identifies occupations involved with the "emerging green economy," and identifies the training needs of a "green workforce."

(HSFL), operated by the University of Hawaii, is now poised to draw new enterprises and technologies related to space launch to the state. Dr. Taylor said that HSFL is preparing to launch and operate small spacecraft from PMRF, which would make it the only university in the world with dedicated rocket launch capability.[7] The HSFL has begun to work in partnership not only with other university and federal laboratory groups, but also with Aerojet Corp. and other private firms in developing new technologies.

While its use of CubeSats[8] and other innovative techniques is at the cutting edge of space engineering, the program is also training a new workforce for Hawaii capable of entering careers in space-related fields and helping to diversify the state's economy. In a parallel development, some of the engineers and technicians who have worked on astronomical facilities for the Mauna Kea and Haleakala sites, as well as for other NASA missions, have begun to use their skills to form spin-off companies based in the state. These companies are now employing local workers in making space instruments for local and mainland clients.

Developing Computing Capacity

While the state is alert to opportunities to build on existing strengths, Dan Goldin, a former NASA Administrator, noted in his presentation that Hawaii needs to develop new capabilities to diversify its economy and advance its technological competence. One area of much-needed growth, he said, is the information technology and computer science sector, where the state lags the national level of skilled people.

Mr. Goldin observed that the growing worldwide glut of data and information presented a major opportunity for Hawaii. At a time when sensors and other sources of data produce information faster than it can be processed, analyzed, or even stored, he said, there is a growing need for expertise in data analytics, network analysis, artificial intelligence, and machine learning.

Outlining Hawaii's advantages in such an effort, he identified the islands' favorable location amid the large and fast-growing markets of the Pacific region. He added that the state has a major advantage in the presence of the Hawaii Open Supercomputing Center, a Department of Defense Supercomputing Resource Center managed by the University of Hawaii. In addition, he noted that Hawaii happens to be immersed in the deep, cold waters of the Pacific Ocean, a nearly infinite and cost-effective source of cooling for the enormous demands of computer server farms.

---

[7]Luke Flynn, HSFL director and Hawaii Institute of Geophysics, and Planetology specialist.

[8]"A CubeSat is a type of miniaturized satellite for space research that usually has a volume of exactly 1 liter (10 cm cube), weighs no more than 1.33 kilograms, and typically uses commercial off-the-shelf electronics components. Beginning in 1999, California Polytechnic State University (Cal Poly) and Stanford University developed the CubeSat specifications to help universities worldwide to perform space science and exploration." As defined in *Wikipedia*. Accessed on July 15, 2011.

To take advantage of this opportunity, Mr. Goldin said, the state needed to pass the 2008 Broadband Plan for Hawaii, which has he said been stymied by "turf disputes, bickering, and lack of will." He noted that Hawaii must also raise the capabilities of its work force. Of the state work force, he said, just 1.4 percent was employed in jobs in the category of math and computer science; the national average for employment in this category was 2.5 percent. Finally, he noted that the state must also create a more innovation-friendly ecosystem. Mr. Goldin noted that he had tried several times to start businesses in Hawaii only to find the business environment to be unreceptive to entrepreneurship.

Growing Innovation Clusters

Barry Johnson of the Economic Development Administration (EDA) described how successful clusters can raise the effectiveness of regional assets, which include not only natural assets (such as geothermal resources, wind, and deep, cold water), but companies, educational institutions, and civic groups. He characterized regional innovation clusters as "geographic concentrations of firms and industries that do business with each other and have common needs for talent, technology, and infrastructure." He added that public-private partnerships among companies, educational institutions, suppliers and customers, federal, local, and state governments, foundations and other non-profit entities, venture capital firms, and financial institutions can help create a climate in which businesses grow along with employment.[9]

Partnering with the Federal Government

In her conference remarks, Ginger Lew of the White House National Economic Council observed that encouraging the development of regional clusters is a priority for the Obama administration. These clusters, she said, are an important mechanism to address the current economic slump and revive competitiveness.

"As a nation," said Ms. Lew, "we need to accelerate the transfer of technology from lab to market. The challenge we face in Hawaii and in many regions of the country is how best to connect innovative entrepreneurs to research institutions and other key partners." She noted that the Obama administration had designed specific policies to encourage cluster formation. The role of the federal government in doing so, she said, was based on knowledge sharing and networking rather than on large investment of resources.

---

[9]For a review of the role that state and federal policies can play to foster innovation clusters, see National Research Council, *Growing Innovation Clusters for American Prosperity: Summary of a Symposium,* Charles W. Wessner, ed., Washington, D.C.: The National Academies Press, 2011. For an assessment of recent economic development policy initiatives, see Junbo Yu and Randall Jackson, "Regional Innovation Clusters: A Critical Review," *Growth and Change,* 42(2), June 2011.

Ms. Lew also highlighted the role that federal partnership programs like the Small Business Innovation Research (SBIR) Program can play. Small companies often struggle for funds to develop new ideas, given that venture capital firms prefer to invest in products that already have revenues and companies with some experience in the marketplace. By providing the "first money" on the basis of a competitive selection process, she said, SBIR can help small innovative firms in Hawaii secure access to early-stage capital. In his conference remarks, Charles Wessner cited a recent study by the National Research Council that described the many contributions of SBIR, noting that over a third of the participants are new to the program each year. Overall, the report concluded that the SBIR program was "sound in concept and effective in operation."[10]

MOVING FORWARD

Recognizing that Hawaii is poised to move quickly to grow its innovation economy, academic, government, and business leaders participating in the conference expressed a willingness to collaborate in overcoming the challenges facing this state, to draw on the advice of national experts, and to listen to the concerns of its indigenous community and other citizens.

In drawing the conference to a close, Dr. Greenwood added a note of caution about Hawaii's ability to accelerate quickly with a new strategy of technological innovation and commercialization. "This is not the first time for much of what we are proposing," she said. The state developed a similar plan 15 years ago when a governor's task force reported that University of Hawaii should be the economic engine for the state and should begin to commercialize some of its research, but that plan was not implemented. "The challenge is to do what we know we have to do," she said, emphasizing the need to act based on the recommendations of the 2011 Hawaii Innovation Council Report. Responding to these remarks, Chuck Gee, a member of the University of Hawaii's Board of Regents noted that the conference had strengthened his confidence in the university's role as a driver for knowledge-based growth. He thanked the conference organizers for "bringing a new sense of excitement as to where we can go from here."

---

[10]National Research Council, *An Assessment of the Small Business Innovation Research Program,* Charles W. Wessner, ed., Washington, D.C.: The National Academies Press, 2008.

II
PROCEEDINGS

DAY 1

Welcome

Howard Carr
Board of Regents
University of Hawaii

Dr. Carr, chair of the Board of Regents at the University of Hawaii, welcomed participants to the symposium titled E Kamakani Noi'i, the Wind That Seeks Knowledge. He noted that the symposium was co-sponsored by the University of Hawaii and the National Academies' Board on Science, Technology, and Economic Policy and on behalf of the University thanked the participants for attending. He then introduced Dr. M.R.C. Greenwood, the president of the University of Hawaii.

M.R.C. Greenwood
University of Hawaii

Dr. Greenwood called the conference "an historic occasion" that is drawing together expertise and national attention to Hawaii's innovation potential. She thanked the local sponsors, which included the Hawaiian Electric Company, Queen's Medical Center, Weinman Innovation Fund, American Savings Bank, HiBeam, Intellisis Corporation, and High Technology Development Corp. She also thanked the Hawaii Business Roundtable for its support, as well as colleagues from the University of Hawaii.

She then introduced her "good friend and colleague," Dr. Mary Good, who, she said, "has been one of those individuals in science and technology policy in

17

the United States who has made it possible for us all to work together across a variety of different fields."

Mary Good
University of Arkansas at Little Rock

Dr. Good, the founding dean of the College of Engineering and Donaghey Professor at the University of Arkansas at Little Rock, said that this conference was one of a series convened by the STEP Board that is focusing on innovation policy among U.S. states. The board's objective, she said, was to observe best practices and gather case studies that might inform federal and state policymakers on best practices to accelerate innovation. During this exercise, the board had realized that much of the innovative work and economic growth in the country is happening at the level of municipalities, states, and regions. The plan of the conference therefore is to tap into the experience of people who had developed policies and programs at the grassroots. The current gathering, she declared, was just such an "ideal mix," representing leaders from the business community, the University of Hawaii, and the executive and legislative branches.

She noted that holding the symposium in Hawaii presented "a marvelous opportunity" to hear about best practices in a state that was "a little bit off the beaten path" and which already had innovation centers making good use of partnerships among academia, the private sector, and the state. "This is essential," she said, "because neither the universities nor the business community can do it by themselves. It really takes a partnership, and it looks as though you have started on a path to make that happen." She thanked all the participants for agreeing to tell the story of one state's economic development initiatives.

Dr. Good then introduced the senior senator from Hawaii, the Honorable Daniel Inouye. She highlighted Senator Inouye's long experience in economic development and described him as "one of the people we could always count on to help with science and technology funding at the federal level." She also said that the Senator embodied for several decades "exactly what we mean to be an entrepreneur: he's able to scope out new ideas, to see opportunities, to bring people together, and to make things happen. These are the critical qualities of an innovator and entrepreneur." Dr. Good added that Senator Inouye was "one of the most effective legislators in the nation and personally one of the most inspiring individuals I have ever met."

Opening Remarks

The Honorable Daniel K. Inouye
United States Senate

Senator Inouye began by praising the assembled speakers, which included "senior government, academic, and business leaders, all of them opinion makers and innovation movers." He welcomed the National Academy of Sciences on behalf of Hawaii and asserted that, as a state, "We have what it takes. We have the intellect, the inquisitiveness, and an entrepreneurial know-how to invent and incubate a knowledge sector in Hawaii." He noted that the state was a leader in astronomy, marine sciences, health, and alternative energy research, drawing from Hawaii's clear atmosphere, clean ocean, mild climate, and diverse ethnic population. The state's major weakness, he added, was "our own modesty and humility, as well as our distance from the research centers of the East Coast." In this respect, he said, the National Academies and other visitors could help by serving as validators, advocates, match-makers, partners, and investors.

"Over these many years," he said, "I've been honored to support the University of Hawaii's research endeavors and to encourage the growth of our technology sector. It is my hope that this conference will encourage a greater intersection and connection between the two." He said that his philosophy had been "rather simple: You build it, and they'll come. You invest, and it will grow."

He offered the example of the Pacific Ocean Science and Technology building, or POST, which the University of Hawaii opened in 1997 at a cost of $48 million, about half of which was covered by federal earmarks, he said. Similarly, in November of 2001 the University took possession of a $54 million Navy-built oceanographic research vessel. "With these two assets," he said, "the University of Hawaii was able to recruit the best scientists and the brightest researchers to build an internationally renowned program. I like to believe this was a key to landing one of the 17 prestigious National Science Foundation Science and

Technology Centers in the area of oceanography. These centers are the gold standard in research."

BUILDING ON EACH SUCCESS

He noted also that the mountains of Haleakala on Maui and Mauna Kea on Hawaii were home to the "world's most prestigious astronomical observatories," representing more than a billion dollars of investments and an annual economic impact of about $150 million. Among the investments were the $50 million Mirror Coating Facility, funded by industry, and the $300 million Advanced Technology Solar Telescope planned for Haleakala. "Each success builds on the last," he said, enticing the world's most acclaimed astronomers to Hawaii. A $1.2 billion, 30-meter telescope is now planned for Mauna Kea. "As I said, you build it, and they'll come." He said that Hawaii had also benefited from investments made in the state's growing technology companies. Many of them, he said, have become increasingly competitive and sustainable.

He closed by touching on the issue of the recent congressional session, during which the Omnibus Spending Bill had been blocked. The primary issue, he said, had been the opposition of many members of Congress to the use of earmarks. He said that in fact spending on earmarks had been reduced since 2006 by more than 75 percent and reformed in significant ways. "We've made earmarks completely transparent," he said. "And the entire Omnibus Spending Bill provided for less than three-quarters of 1 percent of federal discretionary funding."

As an example of the value of some earmarks, he noted that the Maui High-Performance Computing Center was initially funded as an earmark in 1993 to support Air Force activities on Haleakala. "And over the years the University of Hawaii has done a fine job managing this asset," he said, "analyzing mountains of data to gather better space situational awareness for national security. The administration and the Pentagon finally realized this should be part of the defense budget, and it is now a DoD supercomputer. Had we not initiated that valid defense requirement, which would also serve as a critical technology cornerstone in Hawaii, this would not have happened."

The same could be said for the Joint Information Technology Center, he added, which was started by native Hawaiian businesspeople and supported by earmarks. The project provides an electronic means of insuring that all warfighters' medical records follow them from the war theatre through evacuation and ultimately through medical facilities throughout the United States. This system also tracks such medical resources as blood, bandages, and medication in the military inventory. "And this all came from Hawaii," he said. "In 2010, it became part of the Defense Department's annual budget request. Lives were saved and lives will continue to be saved."

Presentation of the
Hawaii Innovation Council Report

Moderator:
M.R.C. Greenwood
University of Hawaii

Dr. Greenwood thanked Senator Inouye and welcomed the members of the Innovation Council, inviting them to join her on the platform for the presentation of their report. She said she would first offer some context for the report and explain why it was important to release it during this meeting on innovation. She described "a clear need" in the minds of science and technology innovation specialists for a changed approach to economic stimulation and innovation—not only in Hawaii, but throughout the country and the world. She said that a summary of this new blueprint provided an appropriate context for this joint meeting with the National Academies Board on Science, Technology, and Economic Policy (STEP).

She acknowledged that while it was not easy to attend meetings on the East Coast from Hawaii, "we are the state closest to Asia, where many of the world's most dynamic economies are located. So we have an opportunity to distinguish Hawaii in the way that East Coast universities distinguish themselves—by being partners with the institutions, organizations, and businesses that will be significant in the future."

Dr. Greenwood began with a brief history of the STEP Board, which, she said, had influenced major policies in the United States over the years. This has included a major study of best practices in public-private partnerships led by Gordon Moore, co-founder of Intel, and continues through the study of state and regional innovation policies. She reviewed additional issues that have been examined by the STEP Board, including a report on U.S. patent policy that has the potential to change significant aspects of patent policy. The STEP Board, she said, is also discussing the impact of copyright policy on innovation in the digital era and is studying the impact of the Bayh-Dole Act and the management of intellectual property at universities. The current study, being led by Mary Good, concerns economic competition in the 21st century, including best practices in state and regional innovation initiatives.

STRENGTHENING EDUCATION, BUILDING A WORK FORCE

Further, she said, some of the work of the STEP Board had been consistent with major objectives of the University of Hawaii (UH). She said that the many programs of the university could be described under three major themes, and that these were relevant equally to community colleges, research universities, the business community, and the legislature. One was the Hawaii Graduation Initiative, which seeks to increase the "educational capital of the state." While it was not the subject of the current meeting, she said, it was one of the university's central objectives and the subject of an education summit.

She said that the second UH objective, "creating a 21st century workforce for a $1 billion research industry," was on the agenda for the meeting. The third objective, she said, was directly related to innovation: "Project Renovate to Innovate," a strategy to renew, rebuild, and advance the local infrastructure.

She then turned to the UH Innovation Council, which she had appointed the previous year for the purpose of delivering a full report on the state's competitiveness. She emphasized that the committee's report was being released that day and had not yet been widely read. "It is the best advice that a group of experts can give us," she said, "and we will now present it, talk about it, and post it on the University of Hawaii Web site where it will be open for comments until Feb[ruary] 1."[1]

She began by introducing the members of the committee who were seated near her on the stage:

• Dr. Carl Bonham, executive director, University of Hawaii Economic Research Organization and associate professor of economics, UH at Mānoa;
• The Honorable Daniel Goldin, president and CEO, Intellisis Corporation; NASA administrator during three successive Presidential administrations;
• Katharine Ku, director, Stanford University Office of Technology Transfer;
• James Lally, partner emeritus, Kleiner Perkins Caulfield & Byers;
• Dr. Brian Taylor, dean, UH at Mānoa School of Ocean and Earth Science and Technology;
• Barry Weinman, chair, University of Hawaii Foundation Board of Trustees, and managing director and co-founder of Allegis Capital, a venture capital firm;

---

[1] A key objective of the Dr. Greenwood was to "continue to contribute positively to the workforce and the economy by appointing a presidential advisory council of experts to study the successes, challenges, and opportunities for a high-value economy in Hawaii. This council will advise on the steps the university should take to create a 21st-century capability for innovation and technology transfer to support a multi-billion dollar industry for Hawaii's research spin-off and related services."

- Dr. Mary Walshok, associate vice chancellor of public programs and dean of extended studies at the University of California at San Diego;
- Dr. Hank Wuh, founder and CEO, Cellular Bioengineering and Skai Ventures, surgeon and entrepreneur.

"We're not alone in trying to reframe and move ahead with what we want to do," she said. "The Council on Competitiveness, one of the influential groups in Washington, said the new normal for effective regional leadership starts with a combination of business, business association leaders, and regional economic developers." She seconded Dr. Good's remark earlier that no single group can do it alone. "We have to have a new connected partnership to make it work. Effective regional leadership requires an ongoing intermediary organization that keeps regionalism or, in our case statism, moving ahead. Innovation is a national priority, which the President has said consistently. This must be accompanied by measures that promote competitive markets, spur entrepreneurism, and catalyze breakthroughs for national priorities."

This thrust was not unlike that of the new administration of Governor Abercrombie, she said, as described in the campaign document "A New Day in Hawaii." A key point of this document is that support should go to the entrepreneurial professor to facilitate innovation and technology transfer, as well as to support premiere education and research projects. "Our state priorities echo national priorities."[2]

RECOMMENDATIONS FROM THE INNOVATION COUNCIL

She said she would summarize the ideas in the report through its "four simple, straightforward recommendations."

The first was to *identify and clarify the status of research as an industry* in the state of Hawaii. The university had grown from having a couple of hundred million dollars in extramural funds to nearly $500 million this year, with some uncertainty caused by the lack of an FY 2011 budget. "That puts the University of Hawaii and the state of Hawaii in the national ranking with respect to our faculty's competitiveness, our institution's competitiveness, and our state's competitiveness for advanced projects," she said. "UH wants to create world-class researchers in emerging or strong fields." The university cannot be strong in all areas, she acknowledged, and needs advice about the best focus. It had gone through a similar exercise almost two decades earlier, when it decided to improve

---

[2]In highlighting the importance of research to the state, the council recommended that "UH put forth a strong recruiting effort to attract world-class researchers in special opportunity areas in which Hawaii has a strategic advantage over anywhere else in the world...such as astronomy, oceanography, and vulcanology. . . . Research is an industry and may become an economic sector in Hawaii, with UH as the R&D engine."

its expertise in the ocean sciences, astronomy, and some engineering fields. This decision, she said, had changed not only the university but also "the lives of people around the world and in the U.S." She urged continued collaboration with other research entities in Hawaii, and on the mainland, and in the Asia-Pacific Economic Cooperation (APEC) region. "We want to use the opportunity of the APEC meeting next year to begin to concretize and develop partnerships with the Asia-APEC area in a much more strong and formalized fashion."

Recommendation two, Dr. Greenwood continued, was to *establish a new way of developing technologies* in the state of Hawaii and with partners on the mainland and the APEC region. Within this recommendation is a specific suggestion that the university pursue a new innovation model around an organization called HiTex, the Hawaii Technology Exchange. This mechanism would reallocate and redefine resources and build more academic-community partnerships. It would also develop programs within the university to foster entrepreneurism and identify a physical location for such an institute.

She then showed a slide depicting the development of a small technology-based company, featuring the familiar valley of death when many firms struggle financially. "This slide is used frequently to show what is wrong with the U.S. approach to moving intellectual property and ideas toward commercialization." Along this spectrum of activities, she said, the nation provides good public support for basic research, but still invests less in basic research as a percent of Gross Domestic Product (GDP) than any other developed nation. For the next stage, of translating the results of research into useful products, there is some public funding available, but private funding is not available in significant amounts as

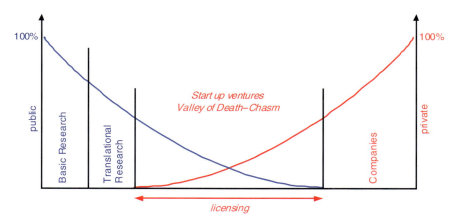

FIGURE 1 Current model of innovation and technology transfer.
SOURCE: M.R.C. Greenwood, Presentation at January 13-14, 2011, National Academies Symposium on "E Kamakani Noi'i (Wind that seeks knowledge)."

a firm enters the critical early stage of commercial development. Private funding begins to be interested only after proof of concept. "So this area in the middle is where start-up ventures need help, in the Valley of Death. I think everyone in the country would like to change that, and here in Hawaii we would like to be a model through an organization that promotes start-up ventures, proof-of-concept centers, regional innovation centers, and this HiTex organization."

A possible structure for HiTex, she said, would begin with an advisory board of not only university people, but also representatives from federal, state, and industry groups, as well as students and post-docs. In addition, it would include several programs called "Come-in-and-come-home," which identify people who have an attachment to Hawaii and who have developed careers elsewhere. "We'd like them to come back," she said, "and develop their ideas here."

Additional strategies would be to include the best practices in intellectual property development and to involve the University of Hawaii Foundation. "The vision is that we get the ideas, we have the workforce, and with HiTex we develop the infrastructure and capital for the future."

Recommendation three, she continued, was to *identify areas for commercialization opportunities*. Topic areas already identified by the committee included (1) security and sustainability, (2) energy and agriculture, and (3) enhancing energy independence and food sufficiency. "We know that these are major goals for the citizens of Hawaii and for the executive and legislative branches of the Hawaii government," she said. Another area, she said, was data analytics. "There's an insatiable need to accumulate and analyze data, and we have some of the largest data sets in the world here in Hawaii. If we were able to master this new and emerging field, we would be a leader, not a follower." Finally, the new UH Cancer Center places it in a position to build expertise in cancers that are prevalent among the populations in Hawaii and the APEC region.

Recommendation four was *to become an entrepreneurial university*: to integrate entrepreneurism into the curriculum, create a fundamental curriculum in entrepreneurism, execute the curriculum, and create an entrepreneur certification process. "As president," she added, "I would like to say to our chancellors and our faculty that I would love the University of Hawaii to be the first university to announce that we will require entrepreneurial experience of every student. We can't do this today, or tomorrow, but we can do it over a period of time. This can change the way we are seen, both in the Asia-APEC region and nationally."

She closed by thanking the committee for its hard work on the report[3] and for its "four concrete recommendations" which would be taken to the university chancellors and faculty and the business community. An implementation plan would be developed, she said, based on feedback from the community and from the current symposium.

---

[3] Available online at <*http://Hawaii.gov/dbedt/innovation*>.

DISCUSSION

In response to a question from an MBA student, Dr. Greenwood noted the report's suggestion that the university should encourage the business school to teach entrepreneurship and also develop introductory entrepreneurial concepts at the community college level and in multiple fields.

Dr. Walshok, a member of the Innovation Council, said that in southern California, digital media artists, storytellers, illustrators, and others used the Web, contributing to "a huge industry peopled by small, entrepreneurial companies." These companies, in turn, found contracting opportunities with the military, the movie industry, corporate training offices, and others. "There's a huge opportunity in the arts and in literature related to digital media arts," she said. "I think the intent of the committee was to emphasize that entrepreneurship is not confined to business schools, but is found in schools of engineering, the sciences, the arts, the social sciences. When you think about new social ventures, demographic trends, and innovative ways to deliver human services, entrepreneurism needs to be everywhere."

Dr. Hinshaw, chancellor of the University of Hawaii at Mānoa, added an example from the arts and humanities department. Students had given a theatre production on sustainability, with the students and audience both learning about the materials they used and the messages they wanted to send. "It does extend to a lot of areas that you might not think of," she said.

Katharine Ku of Stanford University said that the engineering ventures program was "probably the most popular at our university." The program was entrepreneurial, with many start-ups developing from computer science and physical sciences. One "really active" entrepreneurial course, called Bio-Design, tasked medical students with finding a problem, reaching a solution, filing patents, and writing a business plan. She also mentioned a program called SPARK designed to educate faculty on how to be entrepreneurs.

Carl Bonham, of the University of Hawaii at Mānoa, added that the Council talked at length about how best to encourage entrepreneurial activity among faculty members. "One of the goals in high-tech areas is to bring together faculty who have been successful entrepreneurs and have them spread their experience throughout the university," he said.

Daniel Goldin, of Intellisis Corporation, agreed that the common perception of entrepreneurship was limited to high-tech activities from the physical and biological sciences or engineering. "But if one takes a look at this incredible growth of Facebook," he said, "and the whole cottage industry forming around social networking, we see the growth of network analysis and predictive analysis based on human behavior and social skills. In that sense the technologists support the social scientists. So I strongly recommend, as we did in our report, that we support entrepreneurial activity across the whole university, and across all professions. I have a home in Los Angeles where there's a very deep arts community.

There's a spirit of entrepreneurship among the artists and those that support it. We have to look across the board."

Learning to Accept Risk

Hank Wuh, of Cellular Bioengineering, Inc., said that if Americans really believe that innovation is key to our nation's competitive strength, the same holds for the state of Hawaii. "Every city, county, and state in the nation is looking at this issue. There are numerous hubs and creative centers for biotechnology, the life sciences, and social media. But it's really a culture, a mindset. And when it is, young people can feel a sense of fearlessness, that it's okay to take a risk, it's even okay to fail. Coming from an institution of education, that's a critical and difficult message we have to deliver."

Barry Weinman, of Allegis Capital LLC, said that in Silicon Valley, the motto is "Failure is what you learn from." He said that in Hawaii, there is "a bit of a fear of failure. I think that's something the University could help us get past. The venture capital model is ready, fire, aim—because we are ready to take risks before we know what we'll do with whatever we have. When I walk around the campus at UH, I sense a desire for safety. I hear, 'If we're going to transfer technology, let's not do anything that would risk our technology or our grants.' We have to overcome that, or I don't think we will be innovative and commercialize our intellectual property with success."

Brian Taylor, of the University of Hawaii at Mānoa, said that the local mindset about success is shaped partly by traditional federal and other funding sources that expect success. The exception to that is the Defense Advanced Research Projects Agency (DARPA), he said, where they understand that about two-thirds of the ventures they fund will fail. "It's this model of acceptable risk-taking that we have to embed at all levels in the university to make this initiative work."

The Importance of Speed

Mr. Goldin emphasized the importance of entrepreneurship and its close relationship to speed. The university must move beyond its old ways of doing things, he said, which are too slow for entrepreneurship. Many studies of innovation have shown that it does not happen with one individual who has abundant time to ponder. It generally springs up quickly in multiple places at the same time. "So as we go forward here in Hawaii," he said, "we need new ways of doing things quickly without regard to how it fits the old ways of academia. For innovation, four years is unacceptable. It must happen in months rather than years."

Warner Kimo Sutton, an entrepreneur with Diamond Head Renewable Resources, thanked the board for its discussion of entrepreneurship. He added that the process could be accelerated by the presence of a national laboratory in Hawaii, similar to Ames Research Center in California, which would provide a focal point for research and attract more venture capital that was prepared to deal with risk. Dr. Greenwood agreed, and said that the university was in fact moving to enhance some of its partnerships with national facilities, notably the National Renewable Energy Laboratory (NREL) in Colorado.

Susan Yamada, Executive Director of the Pacific Asian Center for Entrepreneurship at the University of Hawaii, voiced her excitement about the Council's recommendation on entrepreneurship. She said that she herself would go even further, asserting the relevance of entrepreneurship "to the individual worker—the student, the employee. We all have to think entrepreneurially; that is the new skill set I believe students need when they're going into the work force—to think innovatively, to see how we can do things more effectively." She noted that even 10 years previously, the university did business in ways very different than today, and it would be still more different in five years. "My question," she said, "having been at the university for a couple years, is, "How do you do entrepreneurship?" You get the big idea, but how can we get to the level of budgeting, of changing mindset and culture? I think that is going to be the biggest obstacle."

Mary Walshok, of the University of California at San Diego, suggested two programs described by the Kauffman Foundation that could serve as models. One was the Deshpande Center for Technological Innovation at the Massachusetts Institute of Technology (MIT), and the other was the Von Liebig Center at UCSD. "What is critical to both of these programs is they engage successful entrepreneurs from the community in the teaching, mentoring, and coaching of students and faculty," she said. "So as Dr. Greenwood, Dr. Goldin, and others have said, it's about a partnership. This is not textbook entrepreneurship teaching—that is doomed to failure. There has to be engagement with real entrepreneurs, and my impression is that there are networks of entrepreneurs across the islands, and there are also very successful entrepreneurs from Hawaii all across the United States, especially in California. They need to be invited into the university, they need to co-create the curriculum and create the opportunities with the students, because they can connect the students to real money, real opportunities, and real markets. And that doesn't cost very much money."

New Criteria for Promotion and Tenure

Dr. Taylor noted that a strength of the university's entrepreneurship center was not only that it invited good speakers and generated an interactive culture but that it also looked system-wide to reach its students. It places students from the community colleges in company internships and in the community. "That's the sort of model that I think works," he said. "I'd say as dean that we've got to change the way we evaluate faculty and how we measure student success. For promotion and tenure, you've got to have entrepreneurship as a criterion. So it's a whole system of things."

Dr. Greenwood agreed that the Council was advising the university not only about how to plant the seeds for the next big industry or set of companies, but also about how to help small business people become more entrepreneurial, to make the family business more successful. "Innovation can be small ideas that work, as well as big ideas that transform an economy."

Mr. Goldin added that in a broader sense innovation should be considered to occur within a kind of ecosystem. He suggested that the university's task was not to be the center of this system, but to help build it. He said he selected San Diego to start his company because that region already had a business ecosystem. He had met most of his employees when they were undergraduates at the university. He brought them into his company, and then some professors naturally followed. The university and the business community, he said, were part of the same "natural ecosystem."

Removing the Regulatory Barriers

Charles Wessner of the National Research Council suggested that if the university intended to take entrepreneurial risks, it should work with legislators to remove regulations that hamper risk-taking. "I think if you want the university to be a research business," he said, "then you need to give it some of the freedoms of a research business." He cited the new College of Nanoscale Science & Engineering near Albany, New York, which had emerged more quickly without the restrictive State University of New York (SUNY) regulations, and the Chalmers University of Technology model in Sweden, where freedom from Ministry of Education restrictions opened the way. The second question, he said, related to the challenge of "feeding" an innovation ecosystem through early-stage financing. "It's one thing to say that you're ready to start a company, but having the idea is not enough. You need private equity in order to grow."

Barry Weinman said he had found a widespread misunderstanding of the term "private" and of the responsibility to use private equity. Private organizations, he said, include not only venture capital funds and companies, but also universities, trust funds, and endowments of many kinds. Such organizations, he said, tend to invest in organizations that support entrepreneurship. In Hawaii, however, he noted the presence of a large capital base that tends to invest primarily on the mainland and could play a meaningful role in promoting Hawaii-based innovation with more of a local focus. He mentioned the example of the Kamehameha Schools,[4] whose endowment is comparable to leading tertiary institutions in the United States, and the employees unions, both of which could contribute to innovation and commercialization in Hawaii. "We have great talent in Hawaii," he said, "and some great research at the university. We can compete

---

[4]Kamehameha Schools, founded in 1887, is supported by a trust created by Princess Bernice Pauahi Bishop, now valued at $6.2 billion. The school system, which supports K-12 and preschool education throughout the state, is the largest independent school system in the United States.

anywhere in the world. But we don't have capital, and that makes it very difficult to build small or help them become national or global."

Hank Wuh echoed Mr. Weinman's comments, adding that even if Hawaii is able to create young entrepreneurs, they will have to "follow the money. What we don't want to do is create a society of great entrepreneurs only to see them leave because there are insufficient resources to support their business." A second essential question, he said, was how to motivate young people for careers in business. At Stanford, where he was a surgical resident, "we had a joke that you can't get tenured unless you've started two companies. That was the whole mindset. You have to make it a badge of honor to be an entrepreneur. They have to be treated like rock stars. And the key is success, so there is a tangible goal for them." Mr. Weinman added that "it's a mental state. We can achieve that in Hawaii, because we have the technology and the talent, but we have to change the thinking."

Session I

The Global Challenge and the Opportunity for Hawaii

Moderator:
Tyrone Taylor
Capital Advisors on Technology

Mr. Taylor complimented the Innovation Council for its work on the report, which he said would "raise the bar for Hawaii's innovation activities," and introduced the speakers for the first panel.

THE INNOVATION IMPERATIVE AND GLOBAL PRACTICES

Charles Wessner
The National Academies

Dr. Wessner, Director of the STEP Board's Program on Technology, Innovation, and Entrepreneurship at the National Academies, praised Dr. Greenwood for her efforts in advancing the innovation dialogue in Hawaii between academic and policy perspectives, and in organizing the current symposium. He introduced the topic of innovation by saying it is "basically about new ideas. I like the definition that says research converts money into knowledge, and innovation converts knowledge back into money." He addressed the common question of why we must focus so sharply on innovation and answered, "We have to. We have to grow our economies. We have to remain competitive, and we have to do that so our children can have a decent future and be secure. Also, the world is not an entirely safe place, and we want our soldiers and airmen to have an unbeatable competitive advantage in their technology."

He said that to meet global challenges, the strongest strategy is to find new technologies and new ways of addressing them. He suggested that this strategy can be visualized in three parts.

The first, he said, is a focus on innovation, which "is how we compete, how we thrive, and ultimately how we win." A new and significant aspect of innovation, he said, is the central role of collaboration. He praised the contributions of Gordon Moore, co-founder of Intel, in leading the work of STEP in the direction of partnerships of government, industry, and university. "He showed us the value of bringing large industries, small industries, universities, and the government together. Those partnerships are what we need to encourage innovation, and they have to reinvent themselves constantly." He cited the work of Dr. Walshok in San Diego as an example.

He said he would talk about global examples of collaboration, some programs in the other states, and "several myths" that have a direct impact on entrepreneurship. The key message, he said, was simple: the rest of the world is more focused on innovation than ever before, and at a high level. He noted with gratification that both senators from Hawaii understood the need for sustained support for innovation. "There's enormous focus here on how you help small and medium enterprises."

Tilting the Playing Field

He added that many economists view the global marketplace as "a place of open competition" and believe that as long as American workers have a level economic playing field, they can out-compete "anyone on the planet." In fact, he said, the playing field is not level. At a fundamental level, he said, the United States begins at an educational disadvantage, as indicated by its low ranking on the PISA studies. In addition, he said, other countries are not interested in a level field; they are interested in winning. "The goal of their policies," he said, "is to tilt the playing field to their advantage."

One way they do this is through strong public support of innovation. He noted that in 1999, China accounted for roughly 6 percent of global R&D expenditures. In 2007, that proportion had risen to 15 percent, and more recent figures show a continued increase. In comparison, he said, the United States makes impressive investments in health research and in defense research. But the amount dedicated to defense research, he said, is misleading, in that much of it—over 90 percent—goes to applied research, including necessary but expensive weapons testing. "These projects are making things work today," he said, "and we need them, but they should not count as research investments for the future. For the warfighter, that means less of a technological edge in the future."

He suggested that the competitive focus of the United States should not be restricted to China, because "the rest of the world is moving as well." He called attention to the success of Brazil's Embraer Company, which manufactures many of the regional jets flown by Americans, of the European partnership Airbus, and the high-speed train network of France, powered by nuclear power plants.

In the case of Brazil, he said, one model to study is its strategy of helping small startup companies to work with larger companies to develop ideas coming out of the universities. The government invests $2 billion a year on this technology transfer, compared with the U.S. Technology Innovation Program at NIST, which has been funded at about $80 million and is perennially threatened with cutbacks.

He also singled out Singapore, with a population of only 4.5 million. "It is not very big," he said, "but it is very determined. They have an incredible focus on innovation, and they're hiring people from all over the world, including Nobel laureates as well as young professors. They have two of the best science and technology parks in the world, Fusionopolis and Biopolis. There is virtually no free land in Singapore, so the determined people build them vertically."

Sustained and Focused Programs

In Europe, he said, countries have sustained and focused programs, many of them drawn from the U.S. experience. "They think some of our programs, like SBIR, work well," he said, "while we have trouble getting them renewed by the Congress." Germany is a high-wage, highly regulated country, like the United States, but nonetheless competes successfully with low-wage economies. It does this through targeted interventions, especially investments in job training, worker retention, and support for small manufacturers. "They understand why manufacturing matters," he said. "They understand, unlike many orthodox economists, that a purely service economy is not equipped to defend itself militarily or employ all the people who need jobs." Germany has focused on green technologies, becoming one of the largest producers of solar technologies in the world, despite receiving no more sunshine than Alaska.[5]

In the United States, he said, the greatest needs are for more speed, critical mass, innovation, and early-stage investments in young, innovative companies. He stressed speed in particular, and the advantages of collaboration between industry and academia. "A 21st-century university is one that partners with industry," he said. As one initiative that had benefited from moving quickly, he cited the success of New York State in deciding boldly and rapidly to build a new nanotechnology center and college. By investing about $2 billion in a single project near Albany, the state attracted more than $10 billion in private investments and succeeded in establishing a world-class center in nanotechnology, including the participation of the SEMATECH program formerly based in Austin, Texas.

He also mentioned a similar success in Michigan, where the economy had suffered severely from the troubles of the automotive industry. To help ensure that the state's auto industry remains competitive in the coming era of electrified vehicles, the state, with active assistance from U.S. Senator Carl Levin, had invested more than $1 billion in grants and tax credits to manufacturers of lithium-ion

---

[5] <http://www.dowsolar.com/why/>.

battery cells and packs. As of mid-2010, some 16 battery-related factories were being built in Michigan, projected to create 62,000 jobs in 5 years.

Universities as Engines of Economic Growth

In Hawaii, he continued, the state had recognized the need to foster, attract, and retain a young and skilled workforce and to support entrepreneurial talent. It had also demonstrated its understanding that "a university is not just a citadel on a hill," but an engine for economic growth. He noted that as a native of Pittsburgh, he had grown up in an era when steel mills were the largest employers in the region. Today, he said, the largest employer is the University of Pittsburgh, which employs not only professors and staff, but also the technicians, gardeners, maintenance people, and many others who are "part of this prospering dynamo that attracts federal dollars." Investments by the university have led to much larger investments by the National Institutes of Health, he said, and those investments in turn "are now resulting in a virtuous cycle where new companies spawned by the research activities are themselves generating new growth."

He said that an important message for the states is that they do not need to "go at it alone." Building innovation success is difficult, he said, but it can be eased by state-federal and state-industry partnerships. He recalled the commitment of President Obama to science and innovation, which he outlined in a major speech at the National Academies soon after taking office.[6] Part of that commitment, he said, is to create the investments and regulatory environment that support innovation. Just as investments in farm crops rise and fall with the level of R&D tax credit and other subsidies, investments in wind, solar, and other renewable energies also depend on the financial environment. "Investments in solar are simply not going to happen by themselves," he said. He recalled the commitment of Congressman Giffords of Arizona to solar technologies, and her understanding that they could not compete against fossil energy sources as long as the level of federal subsidies to the fossil fuels industry remained high. If we don't readjust the policy mix, he said, the countries that have done this, notably China, Brazil, and the European Union (EU) countries, will come to dominate renewables industries. "You don't make these investments without having the resources to make them," he said. "And that applies at the state level as well as the national level."

He emphasized that Hawaii had unique opportunities in energy security, given its natural wind, solar, and geothermal resources. "But you need the collective will and to use your political leverage to initiate the right projects as fast as you can," he said. The key steps, he added, were to arrange partnerships with

---

[6] <*http://www.whitehouse.gov/the_press_office/president-obama-lays-out-strategy-for-american-innovation/*>.

the federal government and to gather support for innovation mechanisms such as S&T parks, research consortia, and innovation awards.

The Myth of Efficient Markets

He returned to the topic of high-tech companies. These are sources of jobs, ideas, and growth, he said, but they face challenges. The most daunting of these is to raise the early funding needed to prove the technical promise and commercial value of a new technology. Another is the "very strong myth" that the commercial markets work efficiently and naturally to meet this need, selecting good new ideas and providing the funding they need as if by an invisible hand. This does not always happen, he said, citing Google as an example of a company that had trouble raising early money. Our federal government invests a lot of money in research, Dr. Wessner noted, but we don't invest enough in the transition from research to commercialization.

Another popular myth, especially among legislators, is that venture capital (VC) companies provide sufficient support for worthy high-tech companies. While the VC markets are broad and deep, he said, and generally considered the strongest in the world, they are seldom a good fit for small companies. VC firms prefer to invest in companies that already have revenues, and preferably profits, and some experience in the marketplace. Moreover, they prefer to invest as late as possible in a company's development, and then to exit as soon as possible—having earned a return on their investment. The total pool of venture funding is only about $1.7 billion for the whole U.S. economy, he said—enough for only 312 deals in 2010. "That is really not a large number," he said. "So while venture is important, it is not the panacea."

So how can small companies get across the "valley of death"? Dr. Wessner urged more strong candidates to take advantage of the federal Small Business Innovation Research (SBIR) program which, along with its smaller cousin the Small Business Technology Transfer (STTR) program, invest nearly $3 billion annually in small firms emerging from university research—nearly double the amount invested by the venture industry. "SBIR gives small awards to begin with," he said, "but that is a strength of the program; $150,000 can attract the attention of a university researcher." For the second stage, the awards rise to as much as $1 million—an amount that had recently been increased by Congress at the recommendation of the STEP Board. "That's an amount that is significant not only to faculty, but to the university leadership."

He added that a further strength of the program is its rigorous and competitive selection process. The National Academies had completed a study under Dr. Jacques Gansler, a former under-secretary for technology at the Department of Defense, which concluded that the SBIR program was "sound in concept and effective in operation." One reason for its effectiveness, said Dr. Wessner, is that

U.S. Venture Capital by Stage of Investment 2009

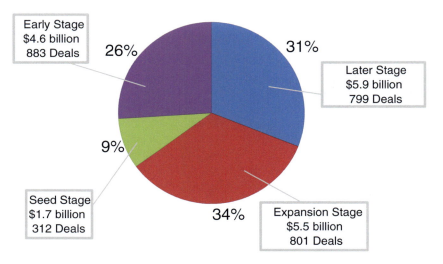

Total: $17.7 billion, 2,795 deals

FIGURE 2 Large U.S. venture capital market is not focused on seed/early-stage firms: U.S. venture investments down 37 percent in 2009.
SOURCE: Charles W. Wessner, Presentation at January 13-14, 2011, National Academies Symposium on "E Kamakani Noi'i (Wind that seeks knowledge)."

it provides "that hardest of all money, the first money, which is needed to get a company started."

He concluded that the report of the UH Innovation Council represented an important first step in accelerating the state's economic growth. He also observed that Hawaii has "a new governor who understands the importance of innovation,"[7] along with a strong and experienced congressional delegation. "Now you need both federal and state investments to help leverage the private investments you need. State investments can be the first critical catalysts in demonstrating the commitment of local institutions, especially your foundations and other investors."

In closing, he encouraged the state to adapt an expression used to powerful effect by a former resident of Hawaii, Barack Obama. "In the case of innovation for Hawaii's growth," concluded Dr. Wessner, "the proper version of President Obama's saying would be 'Yes you can.' I do believe that is true."

---

[7]During his campaign, Governor Neil Abercrombie prepared a roadmap for Hawaii called a "New Day in Hawaii." See *<http://www.neilabercrombie.com/images/uploads/AFG_ANewDayinHawaii.pdf>*.

DISCUSSION

Mr. Weinman reiterated that some of the difficulty experienced by small companies was related to larger issues of private financing. In Europe, he said, the venture capital industry had performed rather poorly because countries lacked strong public markets for private companies. The only growth strategy for a small firm was to be acquired by a larger company, and without the competitiveness of a public market, large companies could acquire smaller companies for lower prices. The United States had been successful in maintaining a strong public market for private companies, he said, until about 10 years ago, when the bursting of the speculative bubble in Internet stocks had led to the passage of the Sarbanes-Oxley bill and other regulations that had depressed this market.[8] Now many small companies have to look abroad for investors, he said, raising the question of how to reduce the regulations imposed in this country.

Dr. Wessner noted that the bursting of the Internet stock bubble and the causes of the recent recession had indeed raised animosity toward parts of the financial sector in the United States. He said that in his personal opinion, the passage of time would mitigate some of those feelings. He also suggested that other factors were weighing on investors. One was a diversion of money from the venture system toward alternative investments that seemed to be more attractive. Another was the challenge of earning sufficient return by investing in small companies during times of economic pressure.

Dr. Lew noted that the Dodd-Frank bill did provide some relief for small companies[9] from certain requirements of Sarbanes-Oxley. She also suggested that there might be an opportunity for some of the larger players, especially the New York Stock Exchange and NASDAQ, to think about establishing alternative exchanges for small-cap companies.

A questioner asked Dr. Wessner what kinds of investments the state might make in support of small businesses. He said that one model for Hawaii would be the strategy of New York State in developing a new nanotechnology educational center with a direct link to industry. He suggested that Hawaii continue to capitalize on some of the unique biological characteristics of its population, including support for the cancer research institute. Another appropriate investment would be a regulatory and perhaps a tax-incentive approach in support of electric vehicles. "A concern about batteries for many states," he said, "is whether

---

[8]The Sarbanes-Oxley Act of 2002 was enacted as a reaction to a number of major corporate and accounting scandals, including those affecting Enron, WorldCom, Tyco, and other companies. These scandals cost investors billions of dollars when share prices collapsed and shook public confidence. While the Act is praised for improving transparency and internal controls, it is criticized for its costs, especially for smaller firms.

[9]The Dodd-Frank Wall Street Reform and Consumer Protection Act was passed in 2010, in the wake of the late-2000s recession. In enacting Dodd-Frank, the Securities and Exchange Commission (SEC) exempted small companies from some of its reporting requirements.

they let you drive 400 or 500 miles. That doesn't seem to be an issue here on the islands. I think you could be seen as an innovation state drawing students, capital, and entrepreneurs from all over the Pacific. You have wonderful assets here, and the task is to make those investments, strengthen the university system, and build more clusters of economic activity that are associated with the university."

Dr. Lew returned to the earlier discussion on investing, stressing the importance of "investing in yourselves." She repeated the observation that major institutions in Hawaii are not making investments in their own economy. She emphasized that she was not calling for giveaways or grants, but investments seeking market-rate returns. As precedent, she cited the cases of Michigan, New York, and California, whose state pension funds are all required to allocate a certain proportion of their investments to in-state companies. Similarly, the New Economy Initiative for Southeast Michigan, a collaborative effort among 20 foundations, has committed itself to revitalizing Detroit by investing $100 million dollars in innovative companies that pledge to remain in Michigan and create high-paying jobs. "These are the types of partnerships and collaborations," she said, "that you might think about."

STATE AND REGIONAL ECONOMIC CONTEXT

Carl Bonham
University of Hawaii Economy Research Organization (UHERO)
University of Hawaii at Mānoa

Dr. Bonham, executive director of the University of Hawaii Economic Research Organization (UHERO) and associate professor of economics at UH, said he would offer his perspective on Hawaii's current prospects as a regional economy.

He began by noting that Hawaii had always been a one- or two-industry state, led until recently by exported natural products.[10] He showed the rise and fall of the economy's major industries, beginning with sandalwood, which ruled from 1806 until 1836, and followed by whaling, which dominated until 1881. This was followed by the "king of agriculture," sugar, which remained the number one export much longer than the others, and shared its leadership toward the end with pineapple. Each lead industry went through a similar cycle of dramatic growth, followed by a peak and fairly rapid decline.

The military security sector quickly rose and just as quickly peaked during World War II, when more than 500,000 U.S. troops were stationed in Hawaii. This number declined slowly over the decades until a federal program of base closures in the 1990s resulted in a loss of 10,000 to 15,000 thousand military personnel. Today about 40,000 troops are stationed in Hawaii, but even at this level

---

[10]Dr. Bonham acknowledged the contributions of Chris Grandy and Bob Shore.

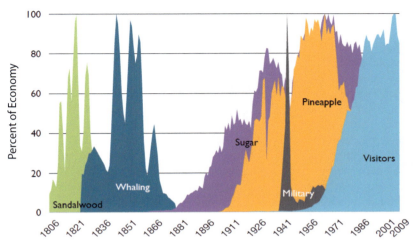

FIGURE 3 Hawaii's predominant industries 1806-2009.
SOURCE: Carl Bonham, Presentation at January 13-14, 2011, National Academies Symposium on "E Kamakani Noi'i (Wind that seeks knowledge)."

the military still accounts for a significant portion of economic support. Federal spending on both military and civilian employment in Hawaii represents more than 12 percent of the economy.

Tourism, which has grown steadily since World War II, has long since taken the lead as the dominant sector, accounting for roughly 15 to 20 percent of the overall economy. It remains highly vulnerable to economic conditions, however, and in the recent recession it declined sharply.

Job Growth in Health Care and Tourism

Overall, said Dr. Bonham, the share of jobs in manufacturing, construction, and agriculture had all declined significantly from 1972 to 2007, "so we're not making much of anything in Hawaii anymore." There has also been an apparent decline in federal government civilian jobs, he said, but in fact the federal employment had remained stable while the overall job numbers have grown. Much of the real growth had taken place in health care and in the service sectors related to tourism. He noted that public and private education were important sources of jobs, but that there were no data for this sector before the 1980s, so its growth was not well documented.

With regard to the standard of living of the people in Hawaii, he showed a graph that compared U.S. real GDP per capita with Hawaii real GDP per capita, indicating that the state had not kept up with the national average. "The long-term trend for productivity growth in Hawaii," he said, "has been roughly 1 percent, while the growth for the United States is close to 2 percent."

Hawaii's uneven growth has been influenced by several exogenous factors. For example, the 1980s showed rising economic growth, but it was caused primarily by three forces: (1) the visitor boom, including many Japanese tourists, (2) a Japanese investment bubble, when capital flowed in to build hotels and buy houses, and (3) state investment in roads and other infrastructure. These forces had the cumulative effect of a boom. "But then we lost a decade," Dr. Bonham said, "that coincided roughly with the 'lost decade' of Japan." This slump was followed by another boom, but today the more important concern is on generating productivity growth that is not dependent on outside forces. "One of the most important ways that we can do that is by insuring that we are essentially creating demand for occupations that are highly productive."

For the period 2000-2009, 27 percent of the occupations in Hawaii were in a category he called "high human capital," in which more than 50 percent of the occupations in the category require at least a bachelor's degree. In most of the

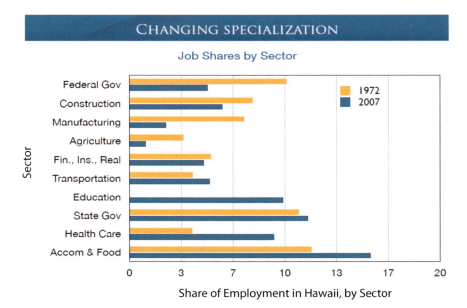

FIGURE 4 Employment share by sector: 1972 and 2007.
SOURCE: Carl Bonham, Presentation at January 13-14, 2011, National Academies Symposium on "E Kamakani Noi'i (Wind that seeks knowledge)."

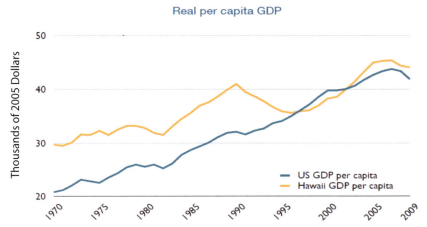

FIGURE 5 Comparing U.S. and Hawaii GDP per capita growth.
SOURCE: Carl Bonham, Presentation at January 13-14, 2011, National Academies Symposium on "E Kamakani Noi'i (Wind that seeks knowledge)."

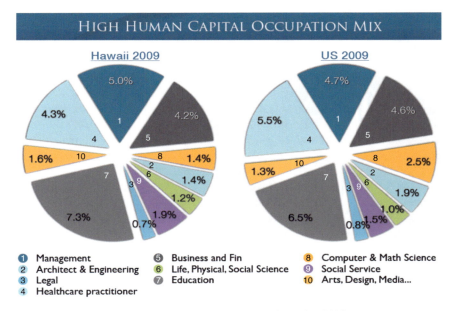

FIGURE 6 Comparing U.S. and Hawaii's human capital mix (2009).
SOURCE: Carl Bonham, Presentation at January 13-14, 2011, National Academies Symposium on "E Kamakani Noi'i (Wind that seeks knowledge)."

tourism-related occupations, however, only about 20 percent of the jobs required a bachelor's degree. In other categories, the percentage requiring a college degree was even smaller: for example, 5 percent in construction, near zero in building and grounds cleaning and agriculture. While the latter categories may require high skills, they are not the kinds of skills that raise productivity or generate spillovers to other areas.

A Need for Spillovers and Entrepreneurial Activity

He broke down the "high human capital" category and compared it to the United States as a whole. In Hawaii, some 1.4 percent of jobs were in architectural engineering; in the United States as a whole, 1.9 percent of jobs were in that category. Some 1.4 percent of jobs were in the computers and mathematics sector, vs. 2.5 percent in the United States as a whole; in some places, such as New York, 10 percent of jobs were in computers and mathematics. "We need to raise those figures—by ensuring that we create demand for occupations that are the most productive and create spillovers. These are the productive, high-wage jobs, directly tied to university R&D. Recent research on economic development has shown a very strong connection between growing R&D dollars at a university and raising high human capital jobs in the region. An active, engaged research university is a necessity, but it still is not sufficient. We also need many people who are working at the university to go beyond research and help create spillovers and entrepreneurial activity in the community."

Dr. Bonham returned to current conditions in the visitor economy and the near-term recovery of Hawaii's economy that seemed to be under way. Before the economic collapse of 2007, he said, visitors were spending about $1 billion a month in Hawaii. This figure fell by about 25 percent. As of August 2010, the state had recovered almost all of that, but costs had gone up, so the recovery was not yet complete. "But tourism is the big mover and the shaker," he said, "and that's what's going to allow the state to invest in the university and move things forward over the next several years."

He offered a brief breakdown of tourist origins, which was changing rapidly. Since 2000 the number of U.S. visitors had increased by about 20 percent, but Japanese visitors had dropped substantially to only about 30 percent of the level of 2000. The number of Canadian visitors is increasing rapidly, as are visitors from Korea, who have increased more than 70 percent in the past year. The prospects for growth from the rest of Asia and from Europe are very strong.

Essentially, he said, Hawaii today is experiencing a continued decline in state and local government jobs, with a forecast growth of total jobs for 2011 at a little over 1 percent. This growth was likely to occur exclusively in the tourism sector, which is UHERO's near-term focus.

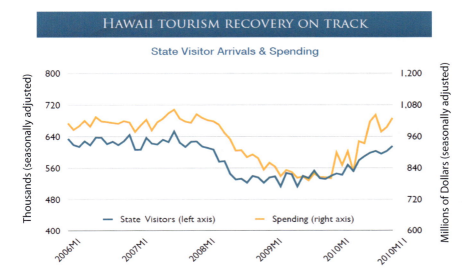

FIGURE 7 Hawaii visitor arrivals and spending 2006-2010.
SOURCE: Carl Bonham, Presentation at January 13-14, 2011, National Academies Symposium on "E Kamakani Noi'i (Wind that seeks knowledge)."

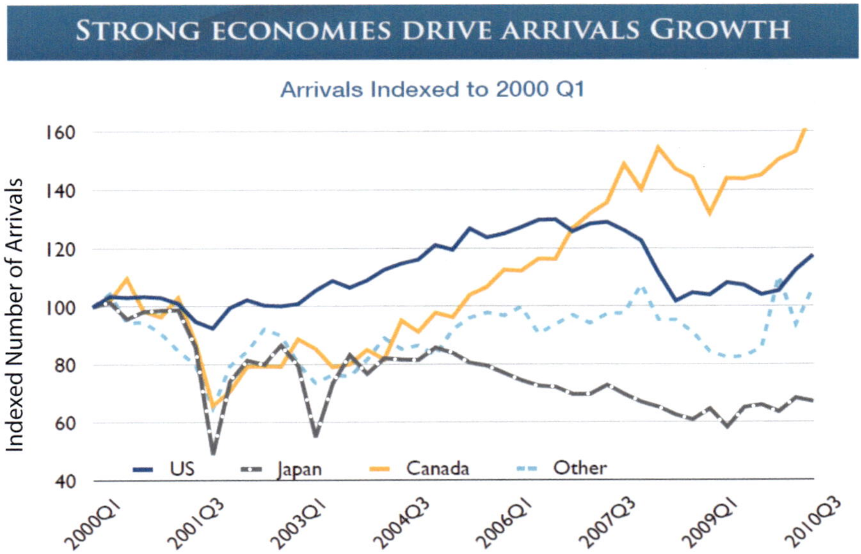

FIGURE 8 Change in composition of international tourism in Hawaii.
SOURCE: Carl Bonham, Presentation at January 13-14, 2011, National Academies Symposium on "E Kamakani Noi'i (Wind that seeks knowledge)."

Investing More in Energy Research

Dr. Bonham asked, how can we move that long-term average of 1 percent closer to the national average? A strategy for doing this, which was essential to raise living standards, begins with the Asia Pacific Economic Community (APEC), he said, and in making connections with other countries and showcasing Hawaii more widely. In addition, Hawaii planned to invest in physical capital, especially in infrastructure that can change and demonstrate how we can produce and deliver energy. "Initially, we're talking about bringing money in from the rest of the world. But the research and development, the focus on this unique place for studying energy, is more important longer term to raise our overall productivity growth. It's the research side much more than laying fiber optic cable or building a solar farm."

He closed by returning to the high human capital group, which was now growing more rapidly. Hawaii actually out-paced the national average growth rate of this sector in the last decade. From 1999 to 2007, this high human capital group expanded by about 20 percent in Hawaii, while the rate for the United States as a whole was 8 to 9 percent.

"That shouldn't be surprising," Dr. Bonham concluded, "because research tells us that when you increase the university's research dollars by 50 percent, as the University has done and is recommended in this report, you are able to bring in the best scientists in the world in areas where we have a competitive advantage, such as astronomy, vulcanology, and renewable energy engineering. When the university grows its research dollars in that manner, it has a direct impact on creating a demand for these occupations. This in turn raises the productivity of the economy and raises the standard of living for all of its citizens."[11]

DISCUSSION

Mr. Goldin commented that of Hawaii's advantages in renewable energy sources, he had not heard geothermal energy discussed, or cooling from deep ocean currents. He suggested that those two resources could meet the enormous power and cooling needs of new cloud computing facilities, now planned for sites on the mainland that lack these natural assets. The Hawaiian coast, he added, could become the centroid for optical fiber cables. "All of a sudden this could spark an information technology revolution for the state. Everyone would want to be here."

Dr. Bonham agreed that such a connection could be very important, as long as uninterrupted, high-quality energy were available. He said that several

---

[11]Dr. Bonham noted that full datasets on the Hawaiian economy are freely available at UHERO's Web site, *<http://uhero.prognoz.com/>*.

colleagues at UHERO were working on sea-water air conditioning and on the energy mix, particularly on the Big Island, where geo-thermal energy can be tapped.

A questioner asked Dr. Bonham about the potential for integrating localized, poly-culture agriculture for both the economic development of Hawaii and to provide additional food security for the islands.

Dr. Bonham agreed that food security was a pressing issue that had not received the attention it merits. He cited a state policy that required identification of important agricultural lands, and UHERO was working on a project on Kaua'i to do that. "But a serious problem for agriculture in Hawaii," he said, "is that if you step back from quality of food and some measure of food security, the highest and best use for land is to build a resort. If you're a large land owner, even if you are farming the land, you're unlikely to invest in improving or conserving it when you can try to get it rezoned and put in a hotel or time shares. As a state, we haven't yet made the commitment to preserve important agricultural lands."

William Harris followed up on Mr. Goldin's comments about capitalizing on local areas of strength, but he questioned the growth potential of astronomy, given the large number of observatories already present. "Despite its excellence," he said, "you may miss opportunities if you don't look to where you need to be rather than to where you are." Dr. Bonham replied that some new growth would come to astronomy as the new telescopes were built and that the university hoped to attract new scientists who want to focus on issues where the state has a comparative advantage. He also agreed that it was essential to move to new opportunities and that the university's objective was not to grow any one science in particular, but to grow the entire research effort.

FOCUSING FEDERAL RESOURCES:
THE OBAMA ADMINISTRATION INNOVATION INITIATIVES

Ginger Lew
White House National Economic Council

Dr. Lew began by thanking the symposium organizers who she praised as "champions of bringing leading researchers and practitioners together to push the innovation agenda." She said she would highlight some of the Obama Administration's policies that promote entrepreneurship and commercialization, especially the efforts to advance regional innovation clusters and improve the commercialization of federally funded research. The two efforts are interrelated, she said, and integral parts of creating durable jobs in the United States.

Dr. Lew turned to regional innovation clusters, which she defined as "geographic concentrations of firms and industries that do business with each other and have common needs for talent, technology, and infrastructure." She went on to note, "Clusters make use of a region's unique assets from academic institutions to workforce to available capital to increased collaboration." Their objective is to create a climate in which businesses can grow and thrive. She said the Obama administration believes that clusters have the potential to create more favorable economic outcomes than conventional models of economic development, including higher wages, increased numbers of business spin-offs, labor jobs that are likely to stay in America, increased regional GDP, a more educated workforce, businesses that are more competitive at home and abroad, and enhanced exports and global trade.

Some Benefits of Innovation Clusters

Dr. Lew emphasized that regional innovation clusters were, however, not a new concept. The United States already had more than 150 recognized clusters, she said, underpinned by a considerable body of research that shows their link to dynamic economic activity. She described one of these in the state of Kansas, which she said is not widely known as a center for innovation or high technology "but makes more than 50 percent of all private aviation technology products and services offered in the United States." Workers employed by the aviation cluster in Kansas earn, on average, incomes more than 50 percent higher than workers not employed in that cluster. She cited a recent study by Michael Porter at Harvard University showing that industries participating in clusters benefit from higher rates of employment and wage growth, more jobs and businesses established, and more patents issued.

During the past year, the Obama administration had made more than 40 awards to support and promote regional innovation clusters. For example, an interagency group had awarded a $129 million award for energy R&D to the Greater Philadelphia Innovation Cluster. This is a consortium of more than 90 public and private organizations, including academic institutions, federal laboratories, industry partners, and federal and regional economic development agencies. This energy research and innovation cluster, or E-RIC, is focused on "developing, expanding, and commercializing energy-efficient building technologies and best practices for national and international deployment."[12]

Working from a Region's Core Assets

But the E-RIC, Dr. Lew said, was only one approach. Her office had now heard from several clusters around the country that were part of emerging or existing cluster initiatives in such diverse industries as alternative energy, health IT, carpet manufacturing, cheese production, and water purification. All of these,

---

[12]For a description of the E-RIC program, see U.S. Department of Energy Web site at <*http://www.energy.gov*>.

she said, were beginning to think more strategically about the core assets in their regions and taking steps to build relationships with each other. In Hawaii, she praised the innovation report commissioned by President Greenwood, which she called a critical first step in bringing together strategic thinkers who could create and catalyze a long-term innovation strategy in this state.

Dr. Lew also mentioned awards made by the administration in other sectors. One was the i-6 Challenge, a competitive, $12 million innovation competition sponsored by the Economic Development Administration in the Department of Commerce, along with the National Science Foundation and the National Institutes of Health. The i-6 Challenge had made six awards in different regions of the United States to promote technology commercialization and new-venture formation. Another program, the Small Business Administration's Regional Cluster Initiative (RIC), had made 12 awards to programs focused on advanced technologies, including advanced defense technologies. Their particular focus was technical assistance, business training, counseling, mentoring, and other services that support job growth and competitiveness in small businesses. The U.S. Department of Agriculture had issued more than 29 awards to different parts of rural America to improve and promote regional economic clusters, particularly in integrated food systems. Finally, the administration has recently created a Task Force for Advancement of Regional Innovation Clusters.

An important feature of this strategy, she said, was that the RIC initiative had broad bi-partisan support in the Congress. At the end of the most recent session, lawmakers from both sides of the aisle institutionalized the RIC as a policy priority in passing the second phase of the America Competes Act. This Act authorized the Secretary of Commerce to establish an RIC program "to encourage and support development of regional innovation strategies," including E-RICs. Through these provisions, the federal government will offer competitive grants and information to stimulate collaborative interactions with firms and other institutions. The government will award competitive grants to the best bottom-up proposals for advancing cluster activities.[13] The legislation encourages regions such as Hawaii to self-organize and also includes incentives for regions to engage the local workforce investment boards and specifically to engage workers displaced by trade. It encourages regions to partner with the private, state, and local groups in both financial and administrative ways. Finally, the Act provides incentives for regional applicants to show how clusters are likely to stimulate innovation and promote economic growth and development.

---

[13]See Section 603 of the America Competes Reauthorization Act of 2010, which establishes a Regional Innovation Program. See <http://www.gpo.gov/fdsys/pkg/BILLS-111hr5116enr/pdf/BILLS-111hr5116enr.pdf>.

Stronger University-Industry Connections

Dr. Lew then turned to commercialization of federally funded research. At the heart of many clusters, she said, is federally funded research conducted at universities and federal labs. These research hubs are integral to a successful innovation ecosystem in enabling researchers and private-sector participants to translate basic research into commercially viable products. A recent study at Harvard University confirmed the essential contribution of universities to innovation, she said, emphasizing the role of the 1980 Bayh-Dole Act in increasing university connections to industry. The study showcased university research, patenting of ideas, and transferred technology. It also documented the economic benefits of research transfer to the local economies through growing employment, higher payrolls, and demand for services that support the university.

Despite this evidence, she cited some recent studies that suggest that America is not doing enough to spur commercialization of its research results. "We spend more on the R side than we do on the D side when it comes to deploying federal dollars," she said. The World Economic Forum had recently issued a Global Competitiveness Report, using 12 key indicators that showed cause for concern: The United States had slipped from first to fourth in commercializing effectiveness. Another recent report by an information technology and innovation foundation called the Atlantic Century showed that the knowledge-based industries of Singapore, Sweden, Luxemburg, Denmark, and South Korea were ahead of the United States in overall competitiveness.

The Obama administration, Dr. Lew said, recognized that commercialization of new technologies is a matter of paramount public interest. The government can exercise tremendous influence over the direction of technology deployment through the $151 billion a year that it invests in research and development, as well as the $76 billion it spends annually on hardware and software. "In an era of limited resources," she said, "we've got to be targeted and strategic in how we deploy our funds. At our national laboratories, we are starting to reevaluate the basic metrics by which we measure success and evaluate how we can accelerate commercialization from our laboratories to the marketplace." At the U.S. Department of Agriculture (USDA), the Agricultural Technology Innovation Partnership (ATIP) programs amplify the work of federal labs by providing proactive, focused, and sustained marketing of laboratory technologies to companies. This can give companies valuable insights into navigating federal processes and accessing resources, such as the SBIR program. Finally, such partnerships enable greater market research to reach the laboratories, thus creating a stronger pull from industry and a better focus for the laboratories wishing to partner with the private sector.

Fostering an Entrepreneurial Mindset

Dr. Lew noted that the Department of Commerce had conducted five regional forums in different parts of the United States with university and business leaders to discuss issues affecting commercialization. In addition, she said that her office and the Office of Science and Technology Policy (OSTP) had issued a request for information that drew more than 200 responses from universities, the private sector, and small businesses, many of them "very thoughtful and rich in commentary." Respondents were asked to focus on several key questions:[14]

• How do universities using federal R&D dollars balance the sometimes competing interests in pursuing knowledge for its own sake and focusing on discoveries that have strong commercial potential?
• What is the best way to integrate universities into broader economic strategies that promote regional economic development?
• How can we foster a more entrepreneurial mindset in universities?
• How can we make it easier to connect entrepreneurs and other "business builders" with ideas generated by university research labs?

In the coming weeks, she said, her office would host several roundtables with private-sector participants from different industry sectors to understand the challenges they face in dealing with universities and research laboratories, especially in licensing and manufacturing products from these facilities. "We recognize that it's not appropriate or feasible to commercialize all federally funded research," she said. "And we also know that different industry sectors have different timelines for commercialization." At the same time, she cited the comment by Dr. Shirley Ann Jackson, president of Rensselaer Polytechnic Institute, that the United States is building today's economy on 20- to 30-year-old technology. "We need to compress that cycle and to accelerate the transfer of our research from the labs to the market. The challenge we face in Hawaii and in many regions of the country is how best to connect innovative entrepreneurs to the research institutions and other key partners."

Dr. Lew praised the plan for developing an intermediary institution in Hawaii and commended other models around the United States. The University of Miami, for example, had placed an entrepreneurship program in its career counseling center. They used this strategy because all undergraduate and graduate students are encouraged to visit that career center, where they could see entrepreneurship identified as a valid career choice. "That means that a music major or science major or business major will all receive the same type of outreach and support through

---

[14]These questions were raised by U.S. Secretary of Commerce Gary Locke in his address to the National Academies R&D Commercialization Forum on February 25, 2010. Access at <*http://www. commerce.gov/news/secretary-speeches/2010/02/24/remarks-rd-commercialization-forum-national-academy-sciences*>.

the career counseling center." Another need, she said, is to provide better access to federal resources for local communities and regions. "We as the federal government need to break through our own silos so that you, the state or the regional community, are not sent around to 35 different agencies to promote one specific objective."

She closed by applauding the "vigorous leadership" she had seen in Hawaii on the part of President Greenwood, Governor Abercrombie, the university, and the "very committed congressional delegation." The symposium itself, Dr. Lew concluded, was clear evidence that "you're starting to think not only about what can you do now, but what you can do in the long term. This is critical. I'd like to commend all of you for your foresight, and I appreciate the privilege to be a part of this important conference."

DISCUSSION

Dr. Harris said he appreciated Dr. Lew's reference to possible new metrics for measuring innovation success. He said that simple measures such as publication counts, patents, and companies were insufficient. "I think we went off track during the post-World War II period when we didn't really compete, because we didn't have any competition. We have to compete now, and we're not doing a very good job at it. I hope you'll begin to think about how you nudge a state such as yours to do things differently." He suggested a pilot program or a federal and state partnership where states, as well as the federal government, each put money on the table. Historically the states have depended on the federal government for research, and "that means the legislators have not paid attention to the workforce," he said, "or to what R&D really means. A few states have made progress, but by and large we have failed to engage local leaders and we have failed to get them the information. So I think you have a huge opportunity."

Dr. Lew said that she agreed with this comment because it emphasizes how many elements there are in the innovation infrastructure, which includes not only roads and bridges, but also education, which "needs tremendous revitalization. I don't think most people realize that almost 30 percent of all Americans do not graduate from high school. Of our population, 18 percent are illiterate, and more than 60 percent can read at only a 7th-grade level. This makes it hard to develop an innovation economy and knowledge-based industries." Investment in education can not be made just from the top down, she said, but has to engage the states and communities "where folks know what the problems are, and where the most innovative solutions can come from."

Luncheon Address

The Honorable Neil Abercrombie
Governor of the State of Hawaii

Dr. Greenwood introduced Dr. Abercrombie as the only governor she knew of who had also been a faculty member—in his case, at the University of Hawaii—and therefore "one of the people who really understands what the university does." He had also been in Congress for 20 years, she noted, providing a rare breadth of academic and political experience in the youngest of the United States.[15]

Governor Abercrombie spoke to the symposium with enthusiasm and optimism, especially in regard to the partnerships described by participants among the university, the business community, and the state and federal governments. He praised the efforts to advance entrepreneurial efforts in the state and underlined his personal commitment to innovation-based growth, as described in detail during his election campaign. He singled out participants from the National Academies, the Department of Defense, and the congressional delegation for understanding the value of dual-use technologies, the SBIR program, and the judicious use of targeted Member Initiatives, or earmarks. "These are actions by a legislative committee in cooperation with the private sector, business, and the university to advance not just individual industries but also the state and the nation," he said. He singled out the 30-meter telescope now planned for Mauna Kea as a reflection of both the hard work of Hawaii's political leaders and an example of the unique resources of the state. "It would be a criminal act not to place that telescope on Mauna Kea," he said. "This is the finest place on earth to explore the heavens."

He vigorously endorsed the recommendations of the UH Innovation Council, which he said overlapped his own campaign issues and the current priorities of

---

[15]Hawaii, the only state composed entirely of islands, was granted statehood on August 21, 1959.

the governor's office. These included especially the Hawaii Graduation Initiative, which seeks to increase the "educational capital of the state," and creation of a "21st century workforce for the research industry." He praised the quality and vision of the university, saying that "everything I hope to be as governor, and in my life, has to do with the university. I came to Hawaii because of the University of Hawaii," where he received "one of the best graduate educations available in the world."

He stressed the value of education for a small state that must rely more on knowledge than it ever has. For Hawaii, which can no longer depend on revenues from sugar and pineapple, the UH can be the new driver to create new businesses in biofuels, geothermal, wind, and other alternative energies; continuing development of technologies of use to the military; and biomedical advances. "We are learning to work together here in Hawaii so we don't have to look to outside sources," he said in closing. "We have the entrepreneurs, the commitment, and the partners to do it. The UH is going to be in the lead in that effort, and I couldn't be happier to make my total and complete commitment to it."

Session II

Leveraging Federal Programs and Investments for Hawaii

Moderator:
The Honorable Brian Schatz
Lieutenant Governor of the State of Hawaii

THE MANUFACTURING EXTENSION PARTNERSHIP: THE NETWORK EFFECT

Roger Kilmer
Manufacturing Extension Partnership Program
National Institute of Standards and Technology

Mr. Kilmer, director of the Hollings Manufacturing Extension Partnership (MEP) at the National Institute of Standards and Technology, began by building on Dr. Wessner's term "real innovation," which "means you're going to make something that somebody's actually going to pay money for. If they don't pay money for it," he said, "it's an invention. I worked at a research laboratory in my previous life at NIST, and I saw that if you don't develop an invention into a product or service somebody will pay for, you're not getting a return on the investment." Helping inventors and small firms capture this return, he said, is the primary focus of the MEP.

The other piece of the MEP name he emphasized was "partnership." Partnership, he said, "really means working together, and if you don't have that, it's more of a coordination role than helping generate real pay-back." He also said that the "next-generation MEP" was focused increasingly on technology and innovation.

The explicit mission of the MEP, Mr. Kilmer said, was to help manufacturers, especially small and medium-sized firms, improve their productivity and competitiveness, and to do this in a strategic way. The program was created in 1988 and has centers in all 50 states, including roughly 370 field offices. The MEP funds about one-third of the operating expenses of the centers, so the centers must be fully integrated with the states and partners in order to function. Revenue for the centers comes in the form of fees for services from manufacturing clients. The MEP interacts with about 34,000 manufacturing firms a year, reaching a "detailed and intensive level" with about 10,000 of them. The 1,400 or so MEP personnel are not federal employees, but are paid through cooperative agreements or grants.

Helping to Think Strategically

The MEP provides practical assistance to help manufacturers to address short-term needs, but it also emphasizes the strategic context of these needs. "Part of the problem we're facing in this country is that we're very short-term focused and reactionary. The MEP tries to promote strategic thinking among manufacturers, help them identify opportunities, and then help find the financial and expert resources needed."

Mr. Kilmer said that the MEP focuses on its impact on clients in evaluating its performance. It evaluates the kinds of services offered to the centers, how well the centers perform, and whether the program gives the centers the right tools and services to help manufacturers. The program regards its work as an investment, so it tries to measure the bottom-line impact, both at the national and local levels. For FY2009, for example, clients reported some $3.5 billion in new sales as a result of MEP assistance, $4.9 billion in retained sales, $1.9 billion in capital investment, $1.3 billion in cost savings, and 72,000 jobs created and retained. These figures were derived from a third-party survey that contacts clients online or by telephone six to nine months after a project is completed.

The primary approach of the MEP is the use of partnerships and networking, at both a national level and the state and local levels. At the national level, Mr. Kilmer said, his job is to link the centers and share resources and solutions for the manufacturers. At the center level, the task is to find the right partners for a manufacturer. This might be a community college to meet training needs, a university to provide research, or the Small Business Association (SBA) to help arrange financing. National partners include federal agencies and trade associations, which are an essential conduit to the manufacturers.

Connections with Local Organizations

The organization as a whole, Mr. Kilmer said, is structured to reach far beyond its federal roots. The MEP has only about 45 federal government employees, most of them stationed in regional centers around the country. However, most of the work is done by third-party suppliers. "We really rely on local people," he said, "because they are our connection with the manufacturing base. We work through a lot of organizations that existed before MEP and are the ones the local clients recognize. It's a unique national asset among federal programs, because we have people who can get to manufacturers anywhere in the country within some reasonable time and effort."

For example, he said, Hawaii's MEP center, directed by Janice Kato, is actually located within the state's High Technology Development Corporation (HTDC), near the University of Hawaii's main campus in Mānoa. The HTDC works with a team of partners including the Department of Business, Economic Development and Tourism; the State of Hawaii Foreign Trade Zone #9; the Hawaii Center for Advanced Transportation Technologies; the Hawaii Strategic Development Corporation; and others.

Moving Beyond Cost Reduction to Product Development

This MEP-HTDC partnership emphasizes product development in the higher-tech areas. "I'll be honest with you," he said, "Hawaii is steps ahead in this emphasis. It's good to see other centers around the country still focused on cost reduction and the lean manufacturing elements, which are necessary. But if a state isn't thinking about growth, as you are, it's probably going the opposite direction. You have to be thinking about the next product, the next service, and how do I get that into the hands of my customers."

From a general manufacturing perspective, he said, trends have been changing, partly due to globalization. He described the MEP's response to three of them:

1. Most manufacturers today are not selling directly to consumers; they sell to another company that integrates the product into something else. Hence manufacturers have to focus on the relevant supply chain and position themselves appropriately. A company does not have to do this alone, he noted; partnerships and business organizations can help a firm develop a new product with high sales potential.

2. The MEP now focuses more sharply on technology adoption by manufacturers, because rates of technology adoption in small firms still lag those of larger firms.

3. A rapidly growing trend is that of sustainability—not only from an environmental or energy conservation perspective, but from the need to build a business model that both satisfies those requirements while helping the bottom line.

In its survey of businesses, the MEP asks about challenges. From a national perspective, he said, companies were focused on controlling costs. In the case of Hawaii, he said approvingly, the top priority was product innovation and development, followed by identifying growth opportunities. "A lot of our effort in dealing with the manufacturing client is getting them to go beyond just cutting costs. We want them to think about how to come up with new ideas and how to get those ideas into products they can take to the marketplace."

Mr. Kilmer said that this emphasis had changed for MEP as well. A decade ago, the program emphasis was on cost reduction; this had given way in the mid-2000s to a focus on strategic management and growth—making companies more competitive, managing growth, and spurring product development. This strategy, in turn, had evolved into efforts to exploit technology to foster innovation. "We really spend the time now trying to get companies to understand that they need (1) to be strategic, and (2) to bring technology into the equation to improve overall competitiveness. It can't be just cut costs, cut costs, cut costs, because that doesn't generate new growth."

Connecting Inventors with Manufacturers

Mr. Kilmer showed a chart he called a "contrast to Valley of Death chart," which he called "the bridge to success." It was meant to show a continuum from research to the marketplace. The MEP's activities were concentrated on the right-hand side, including efforts to improve processes, improve business, and understand markets. On the left-hand side were the people and organizations that generate new research and technology. "For a lot of small manufacturers," he said, "they don't even know what they don't know. That includes the possible markets for a technology, and how they can integrate it into their products or processes. We try to help them reduce the risk of those steps, of bridging that knowledge gap." This does involve financing, he said, which may include private capital, SBIR, or state programs, and it acknowledges that many players and approaches can be involved in the process of commercialization and that all of them need to collaborate to create marketable products. "From an MEP perspective," he said, "we're really that cog that connects the technology folks with the manufacturers. Our focus is on translating the potential of these technologies into business opportunities."

He described a web link called the National Innovation Marketplace, developed in partnership with Eureka Ranch, in Cincinnati, Ohio. Its purpose is to connect those who produce technology with those who manufacture products. He said that it differed from an on-line database in two ways. First, it translates a technology from the language of a scientific paper or patent abstract into the language of business. Second, MEP had done an analysis of market opportunities so that its staff could explain them to manufacturers, including estimates of how a commercial product might be used and how big the market might be. "This helps understand the opportunity and reduce the risk," he said.

Mr. Kilmer said that while the link was intended to connect manufacturers with technology, its effectiveness would rise with its ability to help other members of the value chain. The system now allows manufacturers to post their manufacturing capabilities, which could benefit an original equipment manufacturer

(OEM); it also allows OEMs to post their own needs. By allowing more partners to work together, he said, it could minimize transaction costs while increasing scale. "And we can get this into a structure that my 1,400 field staff can explain," he said, "without knowing the deep engineering or science. They just need to convey in business terms why the manufacturer should make the investment or take the risk."

Helping Manufacturers Diversify the Customer Set

The MEP also works with supply chains so that manufacturers have a better understanding of their core capabilities and how those might be applied more widely. He described a company in Michigan, which had produced gears for a General Motors vehicle that no longer existed. An MEP center there helped the company discover that it could apply the same capability to produce components for a medical prosthetic device. "Now they are diversifying their customer set," he said, "so they're not as reliant on the automotive industry."

Mr. Kilmer said that MEP has worked with other federal agencies and companies in the same way. At the Defense Logistics Agency (DLA) in Columbus, Ohio, the MEP's national network helps companies find manufacturers capable of producing hard-to-find National Stock Number parts. The network also helps BAE Systems, Inc. to communicate electronically with its suppliers for real-time design and production and to help train suppliers. Finally, MEP helped the Veterans' Administration find suppliers that were veteran-owned companies.

A project that grew out of stimulus funding was helping with a statutory requirement to "buy American" in building retrofit programs of the Department of Energy (DoE). The MEP was able to use its network to help the DoE to identify U.S.-based suppliers. As of December 2010, the MEP had made 39 exact or partial matches out of 75 opportunities, and 22 of them were deemed by DoE to be viable U.S. suppliers.

The Complex Demands of Export

Mr. Kilmer described another program, called ExporTech, that helps companies enter and expand into global markets. "Many companies have been used to supplying a customer down the street, or maybe in the next county," he said, "but not in the next country. We try to help them understand what's involved in that and get positioned for the complex demands of export." In the case of Wilco Machine and Fabrication, of Marlow, Oklahoma, which manufactures equipment for the energy industry, MEP had accompanied Wilco officials to visit potential clients in the Middle East and Brazil. After these visits, exports jumped from 8 percent of total revenue in 2008 to 51 percent by mid-2009. ExporTech is a collaborative effort with the U.S. Export Assistance Centers, the SBA, and state and local programs that support export.

Finally, Mr. Kilmer noted that MEP has worked with the Environmental Protection Agency (EPA) for more than eight years to promote manufacturing through sustainability through opportunities that reduce environmental impacts. The strategy is to do this in ways that make good business sense for companies. The MEP supports the Green Suppliers Network and the E3 program on Economy, Energy, and Environment, along with federal partners. The strategy had recently expanded to include energy impacts, financing requirements, and potential training dimensions. This partnership includes the DoE, the Department of Labor, and the SBA. "What's different about the sustainability program," he said, "is that its community based. The partners work directly with local utilities, government, and manufacturers."

He closed by saying that the MEP had worked with many partners over the past half-dozen years in order to maximize its impact and that this strategy had been successful. One program in particular, Mr. Kilmer said, the Interagency Network of Enterprise Assistance Providers, convened with the SBA, had managed to bring together all 82 federal programs that focus on outreach and business assistance and also attracted 36 non-government programs as partners. "We've had such an impact," he concluded, "that we now have organizations outside the federal government that participate in that effort."

DOD STRATEGIC TECHNOLOGY CAPABILITY THRUSTS: OPPORTUNITIES TO FUEL HAWAII'S INNOVATION ECONOMY

Starnes Walker
University of Hawaii

Dr. Walker, chief engineer and technical director at the University of Hawaii, reviewed the technological activities of the Department of Defense (DoD), based on long experience as a science and technology (S&T) director for the Department of Homeland Security, Office of Naval Research, and other positions.

He said he would describe "the programs and thrust areas" for science, technology, research, and development that provide "the capabilities we need to defend the nation, and where we need to go to meet a constantly changing threat."

He began with a set of priorities laid out by Defense Secretary Gates about a year ago: take care of our people, rebalance the military, reform what and how we buy, and support our troops in the field. He said that the activities of science, technology, research, and development "were the seed corn we plant to make those things real. They provide the creativity and capabilities we need across the military."

Dr. Walker began by thanking his colleagues Zachary Lemnios, director of Department of Defense Research and Engineering (DDR&E), and Alan Shaffer, principal deputy, for "providing the focus for the defense agencies in terms of investments and for ensuring collaboration among between organizations." The DoD invests about a $13 billion a year in basic research, applied research, and advanced technology development, all of which "are important to fuel the capabilities we need." A number of studies, he said, had distilled the areas most important in maintaining the DoD's technology capability, beginning with "accelerate delivery of technology capabilities to win the current fight." This "imperative" included electronic warfare, computer science, cyber operations, energy and water, and a rapid capability tool kit. He mentioned especially the importance of electronic warfare. "We are in a cyber connected world," he said. "It's important for us to be able to communicate and to share information. That means we need to have the cyber domain well protected."

He said also that "energy security is national security" and emphasized its importance in the view of Secretary Gates. In addition, alternative energy sources are increasingly important to the DoD as well, because the cost of delivering fuel to troops abroad is "approaching $200 to $300 a gallon."

Another central area, Dr. Walker said, was materials science—the electronic, optical, and physical properties of materials. "These all depend on discoveries that occur across the seams of disciplines," he said. "Physics crossing mathematics, mathematics crossing chemistry, then biology, and now the fertile linkage with human behavioral sciences. We are a network-enabled society and world. So we have to understand how this all fits together."

The Challenge of Moving Science into Practice

A second imperative, preparing for an uncertain future, requires platforms that serve core purposes in battle, such as helicopters, attack vessels, and all-terrain vehicles. Each of them could be further strengthened by new technical capabilities, such as breakthroughs in acoustics, sensors, and materials that provide offensive and defensive capabilities. The challenge, he said, is to both identify important discoveries in science and technology, and then move them into practice. "That's an important area for the DoD," he said, "and something we have to do in a better way." He mentioned some positive programs in this regard, including the Small Business Innovation Research (SBIR) program, such as that managed by the Navy. "I think there are many examples of what has gone well with that, and certainly for the state of Hawaii, we've had some of our greater successes in terms of collective capability in Phase I and Phase II grants." The challenge remains, he said, to "move things from the fundamental discovery stage, planting the thousand flowers, and then harvesting those flowers as something useful."

In addressing the uncertain future, Dr. Walker said, the biggest issue is to decide which challenges to act upon, and to what degree. The DoD has sought advice from outside sources, including the National Academies, the Jasons,[16] and

---

[16]Jason is an independent scientific group, coordinated by the MITRE Group in McLean, Virginia, which advises the U.S. government.

the Defense Science Board to help define the best approaches. The Quadrennial Defense Review Report, compiled every four years, is a core strategic planning document. Many such studies, aimed at "making things real," he said, are public documents accessible on line, where it is possible to see how well the DoD makes use of basic knowledge. Industry, he said, faces the same challenge in "making things real," moving from basic discoveries to useful products. "We have to have a partnership between government, industry, and academia to make this work," he said. "And to make this work we have to build the next generation of outstanding scientists, engineers, and mathematicians." He said that the potential for "technology surprise" is growing because so many more scientists and engineers are now being educated in other countries, so that new research results now appear anywhere around the world. "We need to pay attention to this," he said. "We need to be able to work collectively as researchers."

In response to such changes, Dr. Walker said, DoD strategy had also changed over the past decade. In 1993, he said, the focus was on how the military could conduct two wars simultaneously; there was relatively little concern about non-state actors. "Today we're looking at terrorism and counter insurgency, and also winning the hearts and minds of the people. To me, understanding the human terrain will be an important part of the DoD's investment from now on."

Finding the Capability Gaps

The approach by the military, he said, is to "drill down" to find the capability gaps. "Don't tell me what you want to discover," he said. "Tell me the things that are important to you. Then I will take a look at the investments in basic science that I should be making. And, what are the areas where discoveries might have interesting applications. That's where I will make investments." To show the real utility, he said, the discovery had to be moved into a testing and experimentation environment. "When you make a discovery in the laboratory," he said, it is important to understand it and get user feedback before moving it too quickly into the field.

Dr. Walker returned to the need to operate effectively in cyberspace. The military must defend its systems in a cyber environment contested by nation-states and various sophisticated adversaries. "We need to own that domain," he said. "It's an area that's moving very rapidly. How can we assure ourselves we can connect across the domain and maintain connections?" It is essential to receive secure data from sensor platforms and be able to process it. "How we do that is very important," he said, "because we easily go into sensory overload. We have lots of data, but what does it mean?" This requires extensive fundamental research in the areas of complex adaptive systems and discreet agent models, he said. And we must be able to do sophisticated network analysis of social and behavioral relationships.

Dr. Walker drew a distinction between the physical defeat of capability and the functional defeat of capability. The second kind of defeat shuts down supplies of power, water, fuel, or energy, all of which are connected by cyber and cyber-physical systems. The military's power systems, including pipelines, refining, and other petrochemical operations, are controlled by Supervisory Control and Data Acquisition (SCADA). He recalled the cyber-event that occurred in Estonia over a year previously when the power grid was shut down. "That was done intentionally," he said, "to get someone's attention. We are vulnerable in these areas. We need safe, secure communication, and we have to have it in nanoseconds."

For example, the Navy has electric drives, weapons, and launch capabilities, all of which are connected across all four Navy enterprises: air, surface ships, submarines, and the Navy seals. Each enterprise needs uninterrupted situational awareness of conditions, all cyber-enabled. "I'm here to tell you that the DoD is going to make significant investments across all of these domains," he said, "to ensure that we own that landscape." He noted that one partner in that effort is the new U.S. Cyber Command, established in 2009 in Ft. Meade, Maryland, under General Keith Alexander, who was promoted to four-star rank in 2010. He said that the academic research environment at the University of Hawaii and elsewhere will be critical partners, as will the electronics industry.

In order to maintain state-of-the-art research, he continued, one of the most urgent needs is to provide the next generation of highly skilled scientists and engineers. Dr. Walker said that one of the most urgent concerns of major electronics corporations, including Raytheon, L3, Lockheed Martin, and General Dynamics, is where they will find these trained experts across cyber-related fields. Investment in training is a leading topic of discussion among members of the Business Higher Education Forum, which includes the CEOs of Fortune 500 corporations as members. "They are asking, 'How should industry make its investment in recruiting, retaining, and training the next-generation work force.'"

Systems Must be Able to Absorb Innovation

Taking a broader view of the cyber realm, Dr. Walker said, it is important to be able to "plug and play in an open architecture environment." Systems must be structured so as to absorb discovery and innovation. "You can't have a fixed domain where you make one investment and then you're stuck in one area as technology moves past you," he said. To help achieve this breadth of capabilities, he said the Naval Research Laboratory had invested $160 million a year in funding this area, and was able to leverage this through partner investments from industry and government organizations to produce a capital working fund of some $1.1 billion. "That's an engine of discovery," he said.

He returned to the importance of data in making decisions. "So much of our data now is unstructured, and the meaning is hard to process," he said. "What are the systems we need to process that knowledge?" He listed the research areas of data structure, anomaly detection, embedded algorithms, context, and prediction. "This all comes down to fundamental mathematics. I think everyone in that community has an important contribution to make." In developing new intelligent tools and approaches, he said, he looks at the places where research areas intersect. It is along these "seams," he said, "where we usually find the greatest discoveries."

Partnerships among government, industry, and academia, Dr. Walker said in conclusion, were essential in developing the cross-cutting technologies needed for defense and for civilian uses as well. A key to developing these, he said, was to leverage the R&D work at the DoD, the National Science Foundation, and the national laboratories of the DoE. "And then industry is going to play an important role in making these things real: the processes of information technology, the enabling technologies, and the sensing and data collection."

"I think this conference has been exactly on target," he concluded. "It fits the synergism we expect between industry, government and academia to drive this capability for national security and the capabilities we need."

THE MILITARY AND HIGHER EDUCATION

Vice Admiral Daniel Oliver, USN (Ret.)
Naval Postgraduate School
Monterrey, California

Vice Admiral Oliver, president of the Naval Postgraduate School (NPS), thanked President Greenwood and the National Academies for hosting a symposium on topics "not only important to the health and prosperity of Hawaii, but to the future of the nation." He introduced his talk by saying he would address three issues: the global challenges faced by the nation, a description of the Naval Postgraduate School, and several themes of education and research at the school that may be of common interest.

He said that he would bring a view of the military and higher education that is somewhat different than it might have been during his active duty days. "I was a cold warrior," he said, "and our national security challenges were very different." During the Cold War, he said, the focus was on building capability to win wars. Today, preventing a war is considered as important as winning one. Both 9/11 and the forces of globalization have expanded the national security mission, he said.

"It's a reflection of the understanding that we are all in this together," he said, "and it gives rise to many levels of endeavor and investment of resources. While historically we thought in terms of prevailing in combat, today the span of military missions is much broader." Vice Admiral Oliver referred to a statement by the Chairman of the Joint Chiefs of Staff, Admiral Mike Mullin, who has said that the biggest threat to long-term national security is the economy—not the concern that the United States won't have enough money to buy ships, airplanes, tanks, and helicopters, but the global competition for resources and the struggle between the haves and the have-nots. "All of these things mean that the military interest in research, education, and innovation is broader than I would have visualized even 10 years ago and broader than most people imagine even today."

Vice Admiral Oliver said that in his opinion, current events have been largely defined by the effects of globalization. "Democracy, open markets, and social progress are just a few of the benefits of a global system comprised of mutually interdependent networks of commerce, communications, and governance," he said. "The commons serve as essential conduits through which the global system prospers." Yet recent disruptions at vulnerable transit points, he continued, have highlighted dangers that accompany globalization, which include trans-national terrorism, proliferation of sophisticated weapons, economic instability, and international competition for increasingly vital resources.

Dependence on Vulnerable Systems

"It is no secret that the United States and our partners are facing profound and unsettling shifts in the global balance of power. Ironically, the greatest potential beneficiaries of the global system have in some cases become its most problematic users." For example, he said, anti-satellite missiles may challenge not only military forces but also commercial services and scientific research in space. Recent cyber intrusions have threatened economic and information exchanges, signaling a growing propensity to attack defense and security networks. Increasingly, the U.S. military's freedom of action is threatened by competitors who seek to exploit its dependence upon these vulnerable systems. More specifically, the strategies of potential adversaries seek to deny the U.S. military freedom of action and threaten its ability to project military power.

These disturbing trends, Vice Admiral Oliver said, are especially apparent in the proliferation of advanced air defense systems, solid-fuel ballistic missiles, accurate anti-ship cruise missiles, and sophisticated under-water combat systems. "The challenges before us are complex and daunting," said Vice Admiral Oliver. "In the future we might not know that we are being attacked or by whom, or how to respond. There are many pieces to this global puzzle, and we must as a nation be able to work them all. Hawaii is a piece, and although it is isolated geographically, it is strategically critical to the whole."

Central to addressing such challenges, he said, is higher education for members of the military. At the undergraduate level, the military has its five service academies, but the majority of officers come from civilian institutions where they may join Reserve Officer Training Corps (ROTC) units and prepare for officer candidate schools. At the graduate level, the services have one joint medical school, and each has its own war college, which are highly focused professional schools. For the broader graduate education of officers and military, many are sent to civilian institutions around the world, and others use GI benefits to pursue degrees on their own.

Broad Graduate Education for the Military

Vice Admiral Oliver noted that the only two military research universities that provide a broad graduate education are the Air Force Institute of Technology, in Dayton, Ohio, and the Naval Postgraduate School in Monterey, California. The Naval Postgraduate School, he said, provides graduate-level education for military officers of all services, as well as for government civilians of all federal agencies. It also conducts related research on technology needs for the future security of the United States and its allies. A highly international endeavor, the university normally has faculty working in dozens of countries and currently has 240 international students on campus from 49 countries pursuing masters and Ph.D. degrees.

When the university was founded more than a century ago, it was designed around a core of military-specific engineering programs. Today it has four colleges, 4 major institutes, 14 departments, and some 50 different master's- and Ph.D.-level degrees across science and technology, business, public policy, international studies, and operational sciences. It also offers a range of short courses, certificate programs, and a robust distance-learning program, which accounts for one-third of the student population. The university has a dozen distance learning students in Hawaii, he said, along with multiple research projects.

Of a faculty of more than 700, about 50 are military officers, and about half of the full-time civilian faculty are on tenure tracks. With some 3,000 degree students, the school is fairly small, even though designated as a "global federal service provider." It maintains nearly 900 separate research activities with a reimbursable operating budget that "dwarfs the Navy mission funding for education." It also maintains Collaborative Research and Development Agreements (CRADAs) and other partnering arrangements with well over 100 universities, corporations, federal agencies, and others.

"One thing that makes us unique," Vice Admiral Oliver said, "besides being a federal government organization, is that most of our research assistants are active-duty military men and women with recent combat experience—in many cases motivated by their own experience to find solutions to problems they faced in the field and in the fleet. We also believe that the future of science and technology lies increasingly in collaboration among military operators, industries, civilian academia, and federal agencies. The Naval Postgraduate School and the military look at the upcoming challenges as not just opportunities for expanded partnership in education and research, but as imperatives for securing our homeland in the long term."

In research, he said, his own university partnered with many institutions with similar interests, such as the U.S. Pacific Command and the University of Hawaii. Some areas of common interest that benefit from partnerships, Vice Admiral Oliver said, included the following:

• The future Navy anticipates heavy reliance on autonomous vehicles, particularly in anti-submarine warfare. The Naval Post-Graduate School had been tasked by the Secretary of the Navy to create a dedicated center for this research, which would include field experimentation and student theses.

• The NPS is experimenting with information collection, analysis, and sharing by using commercial devices, like the iPhone, and testing them throughout the Pacific theatre. Many of these capabilities have high potential for dual-use technology spin-offs, such as emergency medicine.

• A critical area for future research is meteorology, oceanography, and ocean acoustics. The U.S. Navy is keenly interested in near-term operational consequences of regional climate change, especially in the western Pacific and the polar regions. NPS researchers are modeling the thinning and recession of the polar ice cap and, through joint monitoring exercises with Pacific Rim partners, deepening the understanding of the dynamics of typhoons and ocean-system modeling.

• A strong research focus for naval strategy is directed energy, such as laser systems. "Lasers may be the ultimate defense against modern anti-ship missiles," he said. "And while some applied research must be closely held, the underlying physics is widely publishable." He said that the school already works closely on directed energy weapon systems and counter measures with Peter Crouch, the University of Hawaii dean of engineering.

• Small "nano-satellites" such as the CubeSat, typically 4 inches on a side, are receiving attention for their potential intelligence value. The goal is to rapidly and cheaply deploy a fleet of CubeSats from a single launcher to provide 3-meter-resolution surveillance. The NPS is a leader in CubeSat research and developed the orbital dispenser in widespread use today.

Energy Goals for the Military

Vice Admiral Oliver noted that energy research requires partnerships and multiple disciplines. "The economic, political, and environmental cost of energy, particularly fossil fuel, is a looming problem for military forces, just as it is for our nation." Middle East oil is a focus of regional instability, he said, so that greening the U.S. Navy is a high priority. The Navy has set "breathtaking" goals for fleets and shore installations, and the NPS has embarked on studies of bio-fuel performance, energy-efficient materials, and fuel cell and battery technologies. The goal of energy independence is common to both the Navy and the state of Hawaii, he noted.

Another area of focus for the NPS is emergency management and preparedness. The Center for Infrastructure Defense has unique analytical capabilities to allow regional and national systems to respond to major disruptions, whether caused by deliberate attacks or accidental events, such as those caused by weather or system failures. The Center is currently working in Hawaii to analyze a commercial fuel-oil distribution system and the inter-island maritime transportation system.

Vice Admiral Oliver summarized by saying that graduate education had become a strategic investment for the military, bringing both short- and long-term benefits to not just military operations but to society as a whole. "I have mentioned some areas in which we at NPS are engaged," he said, "but maybe even more important than what we are exploring today is what we will discover together tomorrow."

INFRASTRUCTURE FOR THE 21ST CENTURY ECONOMY: THE ROLE OF THE ECONOMIC DEVELOPMENT ADMINISTRATOIN

Barry Johnson
Economic Development Administration
U.S. Department of Commerce

Mr. Johnson said he would frame his talk around regional innovation clusters and their role in advancing economic development. His "real theme," however, would be transformation—"because it's transformation that we're all seeking."

Obviously, he said, the nation faces daunting current challenges, including natural disasters, man-made disasters, a housing crisis, the decline of the auto and other manufacturing industries, budget restraints at all levels of government, and persistently high unemployment. And yet, he said, such challenges often bring new opportunities—in this case, to advance the 21st century economy through collaboration, innovation, research and development, higher production with lower consumption, exports into the global market, and the advancements of rising new sectors, such as green technology and low-carbon industries.

A New Model of Economic Development

"The opportunities that await us hold promise for the realization of an economy that works for everyone," he said. "Yet realizing these opportunities requires the adoption of a new approach to economic development policy and practice. We know that the traditional approach to economic development isn't working. Silos don't work. Trying to cultivate jobs without leveraging regional assets doesn't work. And zero or negative sum competition with the next city or county or state does not work. That old model was designed for another time, and a new model is required for a new century."

Mr. Johnson illustrated some of the contrasts between 20th-century economic development and 21st-century economic development in the following terms:

20th Century Economic Development	21st Century Economic Development
Domestic competition	Global competition, collaboration
Zero sum game	Positive sum game
Growth of jobs	Increasing productivity and per capita income
Incentives to attract or retain cost-driven firms & industries	Investments in talent & infrastructure to support innovation-driven clusters
Lead industry attraction and market efforts to firms & industries	Broker innovation networks, connect inventors, financers & transformers
Quantity of jobs, number of firms attracted/ retained	Quality of jobs, wage & income growth, innovation

He noted that while the term cluster had become part of the economic development lexicon, it is used in many different contexts. He offered the EDA's definition:

> *A regional cluster is a geographically bounded, active network of similar, synergistic, or complementary organizations in a sector or industry that leverages the region's unique competitive strengths to create jobs and broaden prosperity.*

Mr. Johnson said that clusters are not defined by their size, or composition, or geography. "They can fit within a political boundary or they can straddle political boundaries, whether those boundaries are counties or states."

More broadly, he said, clusters are based on the presence of regional assets, including companies, educational institutions, suppliers and customers, federal, local, and state governments, foundations and other non-profit entities, venture capital firms, and financial institutions. Each of the various entities has a critical role in supporting the overall cluster, or ecosystem.

A cluster must be inclusive, he said, "and this is where the transformation has to occur. It's not just about sitting around a table; that's just getting together. Collaboration is the shift that occurs when you see yourself as part of the same community as everybody else. You may be the big business with all the money, but you're sitting across the table from a small social entrepreneurship entity or a small non-profit. In this new model, the one with the money doesn't rule. Everyone has a unique role that no one else at that table can play. Each of them needs to be there, and be engaged."

Regional Clusters Across Geographic Boundaries

Mr. Johnson added that a crucial feature of regional clusters is that when they are functioning well, they reflect the function of economies at larger scale. That is, they help clarify economic realities and highlight opportunities where existing geo-political boundaries may present obstacles. For example, while the adjacent areas of Northern Virginia, the District of Columbia, and Maryland are separate political areas, current cluster analysis shows how closely bound the parts of this area are to one another. The activities of industries in the three zones transcend the boundaries between them.

He turned to some of the best practices of clusters, beginning with strong leadership. "The leader may be a business or a university or local government, but leadership has a quality of charisma that demands respect and generates excitement among the partners," he said. Successful clusters are also customized, or focused on unique strengths. He saw this attribute in Hawaii, he said, in its strong partnerships, horizontal and vertical, that included industries, non-profits, foundations, and local communities, both small and large. "Everyone's at the table, ready to make a commitment to collaboration." Another feature of successful clusters is that they understand the markets they need—not just the local markets, but national and global markets. Finally, they have a roadmap for sustainability. "You tap into the broader universe of resources," Mr. Johnson said, "but at some point the federal grant or other support ends. How do you sustain your development?"

Many states, communities, and nations are moving to clusters as a framework for economic development, he said, at least partly because they have proven to yield high returns on investment. The returns have been noted in job creation, business formation, increased competitiveness, increased innovation, and faster commercialization. At least some of these results, he suggested, occur when cluster members attract other members to the cluster.

The Popularity of Clusters Abroad

Mr. Johnson then issued a "red flag wakeup call" in saying that the effectiveness of clusters was no longer a secret among U.S. competitors. "Make no mistake," he said, "our international competitors are fast accelerating their utilization of clusters as a framework for driving economic development. Certainly all the BRIC countries and the EU countries are doing it. They are focused on it, they are funding it, and they are beginning to reap the benefits." He said that the Economic Development Administration (EDA) makes a point of meeting twice a year with about 40 other countries to exchange information about economic practices. "And I could spend an hour just telling you how other countries, both large and small, are embracing clusters."

He noted that the President's economic development policy was rooted in place-based and regional strategies, and that the America Competes Act[17] had described a key role for the EDA in supporting and funding regional clusters.

---

[17]The America COMPETES Act of 2007, which provided a surge of funding for science- and technology-intensive federal agencies, also mandated that each agency cooperate with its partner agencies and offices. The Act was reauthorized in December 2010.

He said that the federal government had four key roles in fostering successful regional economic development economies:

- Advance a common framework: A common regional framework is intended to ensure better alignment of programs and resources.
- Support initiatives: The federal government invests in locally generated ideas and programs deemed capable of spurring regional job growth.
- Convene stakeholders: An important role is to help groups overcome barriers to partnerships. "Many regional players won't think of collaborating for strange reasons," he said, "such as, 'Well, they won the basketball game 42 years ago and they cheated.' We don't have a stake in that argument, so we can come to town and help them overcome those barriers."
- Invest smarter: The federal government has many information resources that are unavailable at the local level, especially about opportunities abroad.

In carrying out its mission, Mr. Johnson said, the EDA does not invest in business directly but in the infrastructure in which business can thrive. "As we make strategic investments to support regional clusters," he said, "we build on a long tradition of best practices to spur regional prosperity. To be sure that the agency is supporting projects that are strategic for the 21st century, EDA supports new investment priorities: projects should be collaborative, demonstrate innovation, support public-private partnerships, push toward a clean-tech, sustainable economy, and promote global competitiveness.

Addressing the Economically Distressed Populations

Mr. Johnson emphasized a final characteristic for today's regional investments: they should address economically distressed and underserved communities. In the past, he said, government has supported some development projects that are meant to help distressed areas, but they fail to do some. Many projects of monumental scale, such as urban renewal projects that replace dilapidated housing with huge buildings, succeeded in displacing rather than empowering underserved areas.[18]

"Too often," he said, "workers would come to their jobs by day and then go home at night to other places. All the people who live in the shadow of this great monument are left unchanged, and still disadvantaged. We believe that's an issue of national security. If we figure out a way to build direct and deliberate bridges into those populations, they can become the talent pool from which our future leaders will emerge.

---

[18]Mr. Johnson referred in particular to the case of Pittsburgh and detailed studies of urban renewal carried out by researchers at Carnegie Mellon University and elsewhere. See, for example, Mindy Thompson Fullilove, MD, *Root Shock: How Tearing Up City Neighborhoods Hurts America, and What We Can Do About It,* New York: Ballantine Books/One World, 2005.

Mr. Johnson noted that the EDA supports a suite of programs that are "essentially investments in the innovation economy." Some of those inputs are data, tools, and technical assistance; some are direct support for projects, including physical improvements of buildings, sewer, roadways, or other infrastructure; or "invisible" networking, such as funds for planning, collaborating, and studying an infrastructure and its strengths and weaknesses. The EDA cluster initiatives include the following:

- Task Force on Advancing Regional Innovation Clusters (TARIC);
- Regional Innovation Acceleration Network (RIAN);
- National geospatial cluster mapping initiative;
- Registry of organized cluster partnerships;
- Urban RICE pilot project;
- Regional Entrepreneurship Action Plans.

Cluster Mapping

"These are the lenses," Mr. Johnson said, "through which we are viewing the investments, and some of them are brand-new." For example, the cluster mapping initiative represents the first such survey. It will be undertaken with Professor Michael Porter and colleagues at Harvard, seeking to understand how the cluster model can best be used and how it can form linkages with other clusters domestically and abroad.

He pointed out that clusters in urban areas have different needs from those in rural areas, so that separate studies have been designed for each. For example, a regional innovation urban project will examine the unique challenges and opportunities of developing clusters in inner cities. "In doing this," he said, "we've discovered a really important principle. That is, the federal government has to transform the way it does business. As it relates to cluster initiatives, that means we need to engage horizontally in a way that we have not done historically, when all the departments worked in their own silos. That's not good enough. Now there's a real effort for the federal agencies to collaborate and reach across those silos."

Mr. Johnson ended his presentation by describing some EDA activities in Hawaii. Over the past four years the EDA has invested $19.5 million as "catalytic money," he said, to leverage more private dollars into projects, including:

- $3 million to renovate a downtown Honolulu warehouse into an import/export small business incubator;
- $3 million in the Maui Economic Development Board to construct a Renewable Energy Resources Center in Kihei; and
- $300,000 to the High-Tech Development Corporation for business start-up assistance.

"We are active and we expect to continue to be active," he concluded. "You're doing the things that align with our investment priorities, so we want to encourage you to keep doing what you're doing and to view us as a worthy partner that can provide guidance and support."

DISCUSSION

Moderator Schatz added several points. First, he said that the established industries in the state of Hawaii were showing signs of recovery since the recession. "We have some good news on the tourism front," he said, "and construction is starting to improve." Second, the Asian Pacific Economic Cooperation meeting was scheduled for Waikiki in November 2011 with 21 member economies. "We think that's a tremendous opportunity for both our public and private sectors to reposition Hawaii vis-à-vis Asia, strengthen existing connections, and rebrand ourselves as a serious place to conduct business travel as well as to conduct business." Finally, he said, Hawaii had now developed an "unprecedented partnership between the University, the federal government, and state government. That's something we used to have 20 or 30 years ago, and it had sort of frayed. But now, with a governor from the University and a university president from the federal government, we have a great opportunity to strengthen these partnerships with the private sector."

Dr. Wessner said he would like to hear more discussion about the issues of industrial policy, which "we on the mainland are dancing around. Every other country in the world has an industrial policy, which really means to make your country or state an attractive place for doing business, and picking strategic areas to emphasize. We've allowed that debate to degenerate into an argument over picking winners or losers, which to me is the wrong debate. And each time we have elections we change policies. The question is, 'How we can make strategic decisions that might last 5 or 10 years? Otherwise it's hard to form a serious public-private partnership, which puts us at a disadvantage relative to Asia and Europe.'"

Mr. Johnson responded that he did not have the answer, but said that all parties needed to move beyond "historic thinking and being. It's about a transformative approach. I find that whenever we advance a policy, even our constituents in economic development tend to ask, 'Why?' We need to articulate the case and demonstrate the upside to show how partnerships are preferred to staying where we are."

Virginia Hinshaw, the UH chancellor, said she noticed that the Manufacturing Extension Partnership in Hawaii did not include the University of Hawaii, and asked Mr. Kilmer how many universities actively participated. Mr. Kilmer replied that he did not know the details for Hawaii, but that universities were the lead partner for operations at about 17 centers. "But partnerships go much beyond that," he said. "It depends on the local region. In the southern part of the U.S. mainland, the universities are the primary economic base from Texas to Florida. In other regions, other organizations may lead, but the universities will still be involved."

Session III

Small Business, Universities, and Regional Growth

Moderator:
Keiki-Pua Dancil
Hawaii Science and Technology Institute

Dr. Dancil, introducing the third session, said that after hearing many case studies and principles from the National Academies and the federal government, the current session would examine how these principles are applied in several parts of the mainland and locally in the state of Hawaii. The current session would lead into the following day's closer focus on Hawaii itself.

40 YEARS OF EXPERIENCE WITH TECHNOLOGY LICENSING

Katharine Ku
Office of Technology Licensing
Stanford University

Ms. Ku said that she was "very ambivalent" about bringing Stanford's experience to the meeting, being "the smallest unit represented from a very dynamic innovation cluster in Silicon Valley." She said she would confine herself to the Stanford experience as an example of what a university can do to make technology transfer effective.

She began with the university's philosophy of technology transfer, which is to do what is best for the technology—to move it into society so that innovations can be useful. In addition, the university sought to foster good industrial relationships. While it naturally had relationships with existing companies, start-ups, alumni, and others, it also seeks good relationships with private corporations, and to do that, it needs to be "reasonable and flexible." This, she said, is best accomplished when there are "very few rules and very few sacred cows."

In addition, the office wants to be action oriented and results oriented. In a university, she acknowledged, these are "often oxymorons," but Stanford had been unusually successful. The technology transfer office was 40 years old, having begun operations in 1970. Since then it had seen some 8,300 cumulative invention disclosures, executed more than 3,500 licenses, and held about 1,200 active licenses. "We have a lot of active licenses because inventions come and go; patents get issued and then expire."

Ms. Ku mentioned a few notable Stanford inventions "just because you probably don't recognize most of them." One of the first was FM sound synthesis, created by a small Yamaha music chip developed by the music department. "We were very proud that one of our biggest inventions, the Yamaha chip, came out of the music department," she said. "It's probably in your cell phone; every time your phone plays some interesting song, that's probably a Yamaha chip."

One of Stanford's biggest inventions was recombinant DNA, the cloning technology that has enabled people to put genes into bacteria. This invention generated some $255 million in royalties, which was shared with the University of California, and licensed to about 440 companies.

"The one that I want to mention just because we're in Hawaii," Ms. Ku said, "is phycobiliproteins proteins. I love this term. It's a fluorescent compound that comes from algae, and it was developed through a collaborative effort between a University of California professor and a Stanford professor. They came up with a compound from algae that can be used in fluorescent-activated cell sorting."

From work in the early 1990s, Stanford researchers also devised a rendition of DSL (digital subscriber line) technology that proved popular. It allowed subscribers a cheaper connection if they were willing to upload slower than they download. In fact, most people do not care how fast they can upload, although they want to download quickly. This technology was eventually acquired by Texas Instruments, which was Stanford's exclusive licensee for many years.

The Success of Google

Stanford's most famous licensee, Ms. Ku said, was Google, a company that was built on a technology appreciated by few people initially. "Two graduate students worked on a project for the library for about four years. They used Stanford resources to develop a search engine that we tried to market to the four big search engine companies, but nobody was interested. The two guys were frustrated, and decided to start their own company. We gave them an exclusive license, but we didn't know if they knew how to do business. We took a little bit of equity, and even that 2 percent share brought in about $337 million in equity. We're happy we were able to give them that start that they needed."

These inventions have generated more than $1.3 billion in cumulative gross royalties over 40 years, with a lot of the money staying with Stanford and the inventors, she said. "And we've been able to give $45 million back to my boss, the

Dean of Research, for the Research Incentive Fund. This money can be somewhat akin to a measure of profit."

The "sort of bad news," Ms. Ku added, is that $870 million of that total came from just three big inventions: recombinant DNA (1974), functional antibodies (1984), and improved hypertext searching (1996). Of some 8,300 inventions seen by the Office of Technology Licensing, only 20 cases had generated $5 million or more, and only 58 had generated $1 million or more. "And I'm always embarrassed to mention this, but we have about $17.6 million of patent expenses in inventory, waiting for somebody to license. Our university doesn't count on royalties for operating expenses, and it can't. The reality is that the big cases almost always end up in litigation, so you never know if you're going to have income tomorrow."

Ms. Ku offered a revenue snapshot of one year, 2010, which was their second best year (after the year of Google). The revenue for 2010 was $65.5 million, from royalties generated by 553 inventions. The lowest royalty amount was about $5, "so we're not always talking about a lot of money." The office spent about $7.1 million in legal expenses for patents; typically it obtains about 80 to 120 new licenses a year, drawn from about 450 inventions a year. One-half to two-thirds are non-exclusive licenses. "We believe in non-exclusive licenses, because we believe that universities should get a fair share when we contribute our technology to a product in the marketplace. About 10 percent of our agreements are with start-ups, so it's not a huge number, about 10 to 12 a year." Some 10 to 20 percent of the licenses are for biological material that researchers use in their research, such as monoclonal antibodies.

She said that with the exception of the Google cash-out, however, equity was not a major factor in the office's revenue scheme. "And I want to tell you," she said, "it costs a lot of money to run the office." It employs 34 people and costs some $5 million a year to operate, not including the $7 to $9 million in patent expenses or the $17 million in unlicensed inventory. Five people work only on agreements with industries that want to sponsor research at the university.

15 Years to Break Even

"The most sobering statistic," Ms. Ku said, "is that it took us 15 years just to break even. The reality is that most of our inventions require a very, very long-term perspective. We have very early-stage inventions. Most of the research inventions are embryonic, so they need lots of years and time and dollars to develop. So we can't ask for maximum royalties or even optimum royalties. Our royalties have to reflect that our technology is very early-stage and very risky. What we're looking for is broad patents, patents that can't be invented around very easily. These have to be revolutionary, not just evolutionary improvements. So that means that most of our inventions take 5 or 10 years before they can come to the marketplace."

Technology transfer is complex, she emphasized. "You generally need a technology champion who really understands the technology to sell it to a company. Patents are only a small part of the picture, but most start-up companies want an exclusive license to give the illusion that they can keep out the big players. Commercialization of university research is very, very high risk, and success really depends on the receptiveness of industry and entrepreneurs."

Several factors affect licensing and technology transfer. The first is the environment, Ms. Ku said. Silicon Valley has a unique ecosystem, where people are "dying to be entrepreneurs. All the graduate students want to know what new start-up you're involved in. It's not just a few people who are interested in entrepreneurial activity, it's almost everybody." Also required is a critical mass of inventions. "Each university, each ecosystem has to decide if it's going to plant seeds or nurture seedlings," she said. "If you're going to nurture seedlings, you want to help the company grow. You may want to offer incubation, or advice and mentorship. Other universities just plant the seeds and let nature take its course. Some of those seeds will blossom and others will die."

The university culture also has to be supportive. President Greenwood and Chancellor Hinshaw, she said, had shown that they are supportive of an entrepreneurial system at UH. The inventors, too, have to be entrepreneurial. It takes a long time to get a patent, to review the patent applications involved, and to take part in the prosecution. "You want your inventors to be involved," she said. "If they're not involved, it's not going to work."

Inflated Expectations

The first challenge for every university licensing office, Ms. Ku said, is high expectations. "You all have a ton of expectations," she said, "I can tell from this conference. But you have to have expectations that match your ecosystem. Does your university have an entrepreneurial culture? The university and the community both have to be risk tolerant. In Silicon Valley, they know that if someone fails, they have learned something and their next venture might be successful. Then, do you have inventors who want to be involved and have the patience to make this happen? Are there companies interested in licensing technologies from the university? And of course, the inventions have to be good. No matter what a university does, you can't sell an invention that doesn't work."

The second general challenge is to balance the many interests involved in technology transfer. "The inventors, the faculty and students, all want us to get more for our technology. The administration wants us not to get them into trouble. The physical sciences companies don't want to pay anything. The life sciences companies recognize patents, and they understand long product development and the value of an exclusive license. The start-up industry wants an exclusive license for the life of the patent, but can't afford to pay much and doesn't know what's going to happen with the technology. Large companies want to control the patent. So there's always tension. "

Ms. Ku observed that the age of a technology transfer office is significant, especially in measuring success. Dollars are not a reliable measure for a young office, she said, because revenues from a particular invention seldom arrive in less than 10 or 15 years. In addition, each deal is different, depending on the company's size, available funding, business experience, and other factors. "We have to be reasonable, mindful of precedence, able to deal with different companies equitably. For a university, one of the biggest challenges is to make sure we honor precedent."

Doing What Is Best for the Technology

Ms. Ku mentioned the National Academies report on managing university intellectual property.[19] She served on the committee and offered two observations derived from her experience. First, technology transfer does fall within the mission of all universities, so the leadership must recognize that. Secondly, "it's not about the money." Given how few inventions return a lot of money, she continued, the main objective of a technology transfer office has to be to "get that technology out to the public."

In conclusion, she offered a broader perspective of the role of her office in the university environment. "We are just a little cog," Ms. Ku said, "and what's important for us is that Stanford maintains its role as a premier university. We want to do what's best for the technology, not chase the dollars. If you're waiting for those dollars, you'll probably miss them, because honestly, we don't know what will be the next big winner. The dollars will come if you do a good job and plant as many seeds as possible. Let the companies find out which ones will bear fruit."

UNIVERSITIES AND ECONOMIC DEVELOPMENT: LESSONS FROM THE "NEW" UNIVERSITY OF AKRON

Luis Proenza
The University of Akron

Dr. Proenza, president of the University of Akron, said he would speak more broadly about the university's role in economic development. He began with a satellite image of the world at night, showing how economic activity, as indicated by the distribution of light, is clustered in geographical regions that bear little

---

[19]National Research Council, *Managing University Intellectual Property in the Public Interest*, Stephen Merrill and Anne-Marie Mazza, eds., Washington, D.C.: The National Academies Press, 2010.

resemblance to mapped geo-political boundaries such as cities and states. In the United States, he said, some 87 percent of economic activity takes place in these larger metropolitan regions. Some 80 percent of the colleges and universities are located there, as are 83 percent of college students. He pointed to a major "blob" of light in northeast Ohio, which crossed many geographic boundaries; this was regional in extent, he said, not local. It contained Cleveland, Akron, Canton, and other economically active places that belonged to larger regions around them. "These regions generally do not compete with each other within the country," he said. "They compete across the world globally."

For each of these individual places, he said, it is important to understand the specific local context and to learn its capabilities. The capabilities of the University of Hawaii were very different from those at Akron, whose local assets included the world's largest program in polymer science and engineering; a strong program in graduate chemistry; a 140-year history as an institution; and leadership in the rubber industry. "What we're seeing today is a convergence among clusters," he said, "because the polymer cluster is now merging with the bio-materials and bio-medical cluster as we begin to develop bio-materials that have functionality and replaceablity in the human body."

Dr. Proenza said that when he arrived at Akron, he found several challenges. Its manufacturing base was in transition from a labor-intensive model to a much more automated model. The region was risk averse, because it had depended for employment on large companies, such as Goodyear and B.F. Goodrich. The universities in the state had no tradition of entrepreneurial activity. Finally, there was little available investment capital.

At the same time, the region had nascent strengths, including underutilized research assets, entrepreneurial potential, opportunities brought by globalization, a convergence of public and private interests, and a focus on differentiation and productivity. To make use of these opportunities, said Dr. Proenza, the university developed a guiding framework. The first principle was relevance, including an effort to focus the university's activities under a strategic vision that utilizes all of its academic disciplines. The second principle was connectivity: the linkages, partnerships, and joint initiatives that bring talents together. Third was the decision to be "exceedingly productive" and to use new metrics to measure this. "We're very interested in measuring the output per unit of input," he said, "so we created scaled metrics to compare ourselves with other larger universities so that we could understand what each actually achieved per dollar of research input."

The University as a Tool Chest

Dr. Proenza looked at the university as a platform or tool chest "that needs to integrate assets within the community through its disciplines and then generate initiatives that benefit the community." The university had designed seven initiatives, each of them based on partnerships among the university, the community, and the private sector. By linking itself with the community, he said, the university could create new value for the community. This was ultimately in the university's own interest, he said, because "if the economy and neighborhoods around us are not healthy," the university itself would eventually suffer the same fate.

The plan began with the concept that the campus itself needed to be transformed. Over the past decade, the university had made a $500 million investment in a "New Landscape for Learning" and created 20 new buildings, 18 major additions and renovations, 34 acres of new green space, 30,000 new trees and bushes, and walkways, plazas, and terraces. "All of this began to have economic impact," he said, "and generate goodwill in the community." At the same time, the university realized that few students were living on the campus, and about 7,000 of them lived in run-down neighborhoods nearby. So a University Park project was initiated, which had grown into the University Park Alliance, revitalizing a 50-block area of about 1,000 acres and 15,000 residents into "a more vibrant and healthy place to live, learn, work, shop, and play." A $10 million catalytic grant by the Knight Foundation to begin the project had led to about 900 new jobs, 80 new houses, $52 million in civic investment, and $300 million in private investment. "Obviously 1,000 acres is going to take a while," he said, "but we are well into stage 2, with the university as developer."

The Value of a Research Foundation

Dr. Proenza noted that the university had also started a research foundation, partly to resolve problems inherent in state laws that tend to restrict entrepreneurial activities of state universities. This issue had been partially resolved in the early 2000s by statutes that permitted faculty to participate in start-up companies, and the foundation had proceeded toward linking industry and the university much more aggressively. As a result, the foundation became a network of entrepreneurial activities that are central to the area's economic development strategies and encompass more than the traditional licensing and commercialization. For example, with a research base of about $50 million, the foundation was able to create about 26 start-up companies based on university-patented technology and another 15 or 20 companies based on patents held by others in a five-year period. He said that compared well with other universities that had a much larger research base.

In addition, the foundation had created partnerships with small and large companies, working with them to develop new technologies jointly. It also developed an innovation campus that serves both to house the research foundation and to provide incubator space. To bridge technology development from concept to actual success, he said, angel investors are needed, so the foundation developed a regional angel network which has also then spawned a student-driven investment network and a women's angel network. As a sideline, he said, they have joined with a community college that had received a private letter ruling from the IRS to have donations for business development grants become tax deductible. "The trick is," he said, "that in order for a company to receive a grant, it has to accept an undergrad or grad student as part of that grant." As a result of this arrangement, which began several years ago, "we are able to improve entrepreneurship across the curriculum, as we call it, with internship opportunities."

The research foundation is also engaged in international ventures, and serves as technology transfer agency for other institutions and organizations. It has also been named the Ohio Research Foundation, rather than the University of Akron Research Foundation, to avoid jealousies.

The foundation had received many awards, including recognition as the most productive foundation in Ohio in the rate of return per research dollar in technology commercialization. It had been recognized by Innovation Associates, the University Economic Development Association, the Milken Institute, and the EDA, which had awarded it one of the six i6 Challenge awards in partnership with the Austen BioInnovation Institute in Akron (ABIA). It had created a partnership with a community college to enhance educational efficiency, to accelerate job formation, and reduce the cost and time required to earn advanced degrees.

Dr. Proenza said that one of the most exciting initiatives of this framework is the Austen BioInnovation Institute in Akron, a $200 million, 10-year program. Akron joined with three hospitals and a regional medical school to create this institute, catalyzed by a significant grant and matched by the state and the private sector. "This is dedicated to becoming the world's number one bio-materials and orthopedic and medical applications research program," he said. "We take the bio-materials excellence at the university, link it with clinical opportunities, and develop the synergies we need. We hold synergy seminars to gather solution providers, or technology owners, with clinicians that have a problem they need to solve. Bringing these groups together has been a very positive step."

He described a new secondary school model based on a partnership with the National Inventors Hall of Fame, which is located in Akron. The university, the city, and the county together created a STEM-intensive middle school to expose students to some of the technologies showcased in the NIHF and "which have transformed the world. We see this as a way to get the students very excited, and we're about to start a high school program as well."

"Guerilla Entrepreneurs"

Finally, Dr. Proenza mentioned a program of the University of Akron Research Foundation that recruits Foundation Fellows, business people who volunteer to act as "guerilla entrepreneurial talent" to identify potential partners for the university. "Ten years ago we would never have thought of partnering with the hospitals," he said, "because we didn't have anything in common at the time. Now these Fellows have helped us to see ourselves as partners with the city and community in a broader sense. As such we can coordinate with other regional partners to expand the concept of the university beyond education, research, and technology transfer. This platform, he said, can focus on relevance, connectivity, and productivity, recognize and resolve the problem of egos, address "relationship fatigue," and let go of short-term control to gain long-term advantage. "This is hard work," he said. "It's a contact sport, takes a lot of time, and requires breaking down a lot of silos." He noted that the university had literally broken down some grain silos originally owned by the Quaker Oats Company; the silos are now transformed into a residence hall and hotel. "We need to do that with other organizations," he said. "We are "silo busters."

Dr. Proenza closed by defining the "new" University of Akron as a "platform" or "tool chest" that allowed a great expansion of the role and potential of the university. "The model," he said, "is that we are no longer alone, but a key partner in the knowledge economy. We are a convener, a developer, an anchor for clusters of innovation. We will not measure ourselves by how many students we exclude, but by the value we add to students and the relationships we build with our communities in solving real problems."

CONVERTING UNIVERSITY RESEARCH INTO
START-UP COMPANIES

Barry Weinman
Allegis Capital LLC

Mr. Weinman, who chairs the University of Hawaii Foundation, began by revisiting the purpose of technology commercialization. He agreed with Ms. Ku that moving new technologies into the market place is not done solely to make money, but said that "it should be a little about the money." He pointed out the large and growing needs of the public universities as state support dwindled for most of them. "When I first came to Hawaii," he said, "the legislature made up 100 percent of the university's budget. I think last year that figure was around 40-42 percent, and for the Big Ten universities it averages only 7 percent. So the university is going to have to find money, and its research output should be one place to find it."

He then turned to the need for improving the way technology is commercialized in Hawaii. "Normally in the venture capital business we are focused on outcome," he said. "We don't care about process very much. On the President's Council, we talked a lot about strategic planning for the outcomes we hoped for. Personally, I often find discussions of process confusing, but I will say that if we don't fix the process, the strategic plan probably won't come to pass. We need everybody's help to fix it."

A "Radical View": Support Universities That Commercialize

The premise of commercialization, Mr. Weinman said, is that research grants at universities create some leading edge intellectual property (IP), and that IP can benefit society in health, business, quality of life, economic activity, and other ways. "I contend," he said, "that over time universities that fail to commercialize their IP may lose some of their future grants. If I were running the U.S. government, I would give money to universities that did commercialize their IP, and put it into the public domain to benefit society. That's probably a fairly radical view, but I think that's the way the public universities will win or lose."

He said that there are many "urban legends" at universities, including the UH, about why it is difficult to convert university IP into start-up companies. "Some examples of these legends are: we've always done it that way; we've never done it that way; the legislature won't give us permission to create a technology transfer office, or to privatize the technology transfer process; maybe the unions will be against it; it's not the faculty's job; we're not Silicon Valley; what happens if we fail?"

"I take the approach that we ought to go and do it," he said. "It's better to apologize later than to wait for permission now."

Mr. Weinman said that Hawaii has been a very innovative place, with traditional skills dating from antiquity when the Polynesian navigators learned to explored the Pacific. More recently, many new high-tech companies have affiliation with Hawaii, "and we don't give ourselves credit for that." The firm Digital Island, he noted, was a local start-up that had a market capitalization of $1.6 billion when it went public. Veriphone, which raised $1.2 billion at start-up, was founded in Hawaii, bought by Hewlett Packard, and then taken public again. "I contend the right thing to do is create wealth, because wealth then gives you the ability to start more companies, invest in yourself, and convince people and institutions in Hawaii, who invest significant amounts of money in other places, to invest here."

Hawaii's New Companies

Mr. Weinman said that a number of new and interesting companies had come out of the UH, including Hoana, which had developed a wireless, noninvasive technology to monitor vital signs. Originally a dual-use project supported by military funding, it was already used in 22 hospitals. Another was Protekai, or Proteins from the Sea, funded by the UH Foundation to market certain proteins from jellyfish. Other young firms included Eyegenix, which makes artificial corneas marketed in Asia where organ donation is rare; TruTags, which makes silica microtags to track medications used in clinical trials; and Adama Materials, which manufactures nano-enhanced epoxy resins and pre-impregnated composite systems. Adama was the winner of the UH business plan competition and had recently raised about $4.5 million from venture capitalists. Finally, Sentilent, a project that the Foundation was trying to commercialize, had developed a technology to determine sentiment and identify opinion leaders by analyzing blogs, Tweets, and Facebook entries. Its work had been supported by $3.4 million from the Office of Naval Research, the U.S. Air Force, and DARPA grants to university Professor Sun-Ki Chai. While many firms are attempting to exploit social networking phenomena, he said, Hawaii has an advantage in its unique cultural mix, along with local capabilities in computer science. "We need to leverage our 'unfair advantages'," Mr. Weinman said, "like jellyfish, ocean water, wind, geothermal heat. If we do that we can create interesting companies."

He said that the UH Foundation now had its own small "seed venture capital" fund, called the Upside Fund. In 2010 Mr. Weinman said that he had been asked to manage this fund, beginning with only $385,000, which made it "the hardest thing I've ever tried to do." It had just made its first investment, and he said several more were planned. "We'll put our money into UH researchers who have an entrepreneurship bent," he said. "We have to convince people to look not only at the market in Hawaii, but also the national and global markets. You have to build the partnerships and position your company with the right management teams to raise money on a global basis. That's something we still don't do well enough here."

Privatizing Tech Transfer

Mr. Weinman proposed a fundamental change in the process of commercialization. "I think we have to take the whole tech transfer process out of the university and privatize it," he said. "Other universities are doing this." He said that one of the most successful was the Wisconsin Alumni Research Foundation, which had brought about $1.2 billion into the university; he said that Arizona State University and the University of Arizona were other examples.

By creating a 501(c)3 corporation, he said, "you can act the way a private company would act, as opposed to the way a university would act." He emphasized that when a product is ready for patenting, the patenting organization must be able to make a decision and move. "We should never be in a position where we can't support our researchers when they want to file something that has merit. Other people are doing this, so why can't we be a leader? The benefits are to diversify Hawaii's economy and hold on to the best and brightest people." Today, he said, these people leave the state because they don't find high-tech jobs.

"We need to be able to create these companies for the next generation. At UH we get major funding for ocean sciences, health sciences, astronomy, life sciences, clean energy. These are our 'unfair advantages.' We ought to focus on these advantages, identify the important innovations, and use venture-type funding to develop the technologies that come out of them."

In practice, Mr. Weinman said, the university's commercialization process needed to be more proactive. Instead of waiting for researchers to ask for help filing a patent, he said, the chief commercialization officer should be "way out ahead of that," meeting with innovators to see where they are going, helping to evaluate technologies, and perhaps bringing faculty or students from the business school to study the market and the competition.

"If IP is unique," he concluded, "we ought to put all our oars behind it and pull as hard as we can. We can help determine whether to apply for a license, spin out a company, join with a larger company, or raise more financing. Then we'll do whatever is necessary to protect it and nurture it. And let's not just focus on the licensing and royalty; let's focus on having successful entrepreneurs who have a warm spot in their heart not only for Hawaii, but for the University of Hawaii. Grateful entrepreneurs are the ones who give back."

IMPROVING INDUSTRY PARTNERSHIPS

Mary Walshok
University of California at San Diego

Dr. Walshok, associate vice chancellor of public programs and dean of extended studies at the University of California at San Diego, spoke about the importance of the community in strengthening university research and research transfer. She said she would illustrate how this can happen through her experience in San Diego, a city and region that in the 1950s faced challenges similar to those of Hawaii—a need to diversify and to transform the regional economy. "But unlike Hawaii," she said, "it had few of the assets one associates with technology hubs today." It had, for example, virtually none of the following assets:

- Land or facilities dedicated to research and education;
- Large, competitive basic research institutions;
- Patenting and licensing;
- Angel or venture capital;
- Technology entrepreneurs or start-up know-how;
- Access to global partners and markets.

Seventy-five percent of the economy was related to the military. And after World War 2, the military-based economy began to shrink as its defense contracting industries lost momentum. In 1962, *Time* magazine described San Diego as "Bust Town, U.S.A. "So you will understand why I cried when my husband told me he was taking a job in San Diego."

Today, some 50 years later, Dr. Walshok said, San Diego is a city of 1.3 million people that has become a major global science and technology hub. UCSD has more than 29,000 students, more than 1,000 faculty, and ranks sixth in research funding in the United States. It is also ranked #13 globally by the Shanghai Jiao Tong University ranking (2008). The once-barren Torrey Pines Mesa is home to 74 research centers, which last year received more than $3 billion in basic research funding and file about 7,000 patent requests per year.

The rapid growth of research and research-based firms began around a core of outstanding research institutions: Scripps Institution of Oceanography (1903), General Atomics (1955), University of California at San Diego (1960), and the Salk Institute (1960). Today there are more than 50 research institutes, with the newer Sanford-Burnham Medical Research Institute one of the largest. These institutes have research budgets of about $100 million to $400 million, in addition to the $1 billion awarded to UCSD last year.

San Diego as a Hub for Technology Companies

"We've become a hub of diverse technology companies," Dr. Walshok said, "growing out of a handful of pioneers. When I moved to San Diego and eventually took a job at UCSD, there were two interesting companies in the category of IT and software. One, Linkabit, was started by a faculty member, Irwin Jacobs, in 1968, based on defense contracts. This grew into Qualcomm." The other was a computer graphics company, ISSCO, started in 1970 by Peter Preuss, then a graduate student at UCSD and later a national pioneer in the field of computer graphics. These isolated entrepreneurial enterprises have grown into nearly 1,000 companies. In the life sciences, the first successful company was Hybritech, founded in 1978 by faculty members Ivor Royston and Howard Birndorf. That company, sold to Eli Lilly, has been joined by over 600 life science companies, anchored by Biogen Idec, Gen-Probe, and Life Technologies.

In addition, another cluster of companies has grown out of energy and environmental innovations. General Dynamics, a major military contractor during World War II and the builder of early nuclear submarines, created in 1955 a new division GA to explore nuclear applications. They recruited major physicists and engineers from across the nation. Today more than 250 companies work in the energy R&D space, all contiguous to General Atomics.

Another booming field, action sports, traces its origins to the founding of Gordon and Smith Surfboards in 1959. Today some 600 companies work in the research and development areas of golf, surfboarding, underwater kinetics, helmets, skateboards, and related areas. "It's an industry," Dr. Walshok said, "and it's based on materials science, developments in physics, and visualization techniques that can be studied in our supercomputer center."

Sources of Success

What brought about 3,000 companies and 70,000 jobs to the San Diego area, successfully replacing many of the jobs lost in the defense downturn of the

1980s, asked Dr. Walshok. Answering this question, she described the interplay of six critical factors:

1. **Land use and infrastructure**. In San Diego, a few civic leaders managed to convince the city council that the only way to keep military activities in the area was to support R&D. In order to do this, land was needed for R&D companies. The council designated a large parcel of "pueblo land," which was land the city had received from the state from the redistribution of Spanish land grants at the end of the Mexican American War in the 1850s. "In our case, these land use decisions allowed for a geographic proximity that is to die for. I only need to walk a mile from my office to pass $3 billion worth of research institutions—not counting the smaller, innovative R&D companies."

2. **World-class research**. The major institutions—including Salk, UCSD, Sanford Burnham, and Scripps—all used the strategy of hiring senior-level people to accelerate research excellence. "When you hire a senior-level person in science," she said, "you hire a person who has connections with the Washington establishment and foundations, and who brings grants, grad students, and postdocs to your region. So you accelerate everything." The university did this by bundling two assistant professor salaries to hire a full professor, then asking the private sector for help in financing, she said. "It became an extraordinary strategy that paid off."

3. **Private-sector investment**. There is a lot of money in Hawaii, Dr. Walshok noted, "but you can't shake it loose." In San Diego the money was simply not there. To catalyze the necessary activities, groups of people contributed sums of a few thousand dollars to start small programs. Most of the organizations that have enabled the growth of the San Diego economy, she said, are built on the efforts of private-sector members, sponsors, and underwriters, and very few federal earmarks.

4. **An entrepreneurial culture**. The culture itself became entrepreneurial, involving "shared agenda setting, shared investment, shared risk, and shared rewards." This culture, she added, was spurred on by crises in the California economy, when no state money was available. The physical proximity of institutions was also an advantage, leading to many formal and informal interactions.

5. **Talent development**. The universities are essential, Dr. Walshok said, in developing the skills needed by emerging technology companies in the region. This may involve special needs at times, when university flexibility can be a crucial advantage. She cited the example of Irwin Jacobs, founder of Qualcomm, who needed engineers trained in his new wireless technology, CDMA, which was not being taught at other engineering schools. UCSD complied, and the company succeeded. "This talent development commitment by the university and the community colleges was an essential component. It allowed us to keep jobs in San Diego and attract people who could develop skills further."

6. **A commitment to place**. Finally, she said, the people of Hawaii and California share a strong commitment to place, which can bring a great advantage.

"Talented people will stay if you give them reasons to stay." Entrepreneurs will mentor young people and become serial entrepreneurs and invest their money in 2nd and 3rd and 8th-and 10th-generation families. They can also become enormously philanthropic. "These characteristics of my place I think resonate with the characteristics of your place," she said. "But what we did over 40 years, you can probably do in 15 or 20."

Dr. Walshok concluded with a recommendation from her own professional field. "I am a sociologist, not an economist," she said, "and before wealth can be created, human beings have to learn to work together. If there is to be progress, new forms of association have to be developed. Part of what the Council is recommending is that the University of Hawaii can be a catalyst and a hub for some of those new forms of association. Economic growth is connected to technological development, but you also need organizational innovation and cultural change. It's been true throughout history, and it's true today."

DISCUSSION

Dr. Wessner asked Katharine Ku whether Stanford innovators used the SBIR program much. She said that they did not, partly because of worries about conflict of interest. "We feel that if money goes to a faculty company and then comes back to Stanford, there could be a perception of pipelining to the faculty company."

Dr. Wessner also asked about the critiques by the Kaufmann Foundation of university technology transfer offices. Ms. Ku said that the performance of various university offices was mixed. "Stanford has a lot of inventions and experience," she said. "Some university technology transfer offices have one or two people, 25 inventions a year, and minimal budgets. It's hard for them to get the experience and then to be hooked to the larger community." She referred to the report on patenting by the National Academies, which recommended that the smaller universities might ask other universities to help them manage their technologies or provide guidance.[20] "And I feel that's part of our role at Stanford—to spread the best practices."

Finally Dr. Wessner asked why universities worked so hard on technology transfer if it was so difficult. Ms. Ku said that Stanford saw it as "part of the mission of the university. This is a public service. We don't want the technologies to lie fallow in the literature. I think we need the inventors to be championing the technology, and we want to help them."

Dr. Proenza added that many universities are doing good jobs at technology transfer, and suggested that the Kaufmann team "had seen many of the less

---

[20]National Research Council, *Managing University Intellectual Property in the Public Interest*, *op. cit.*

optimal offices." He said that the Kaufmann team had been to Akron, and "understood what could be done by this highly robust model."

Mr. Weinman added that university leadership could also play a significant role in moving technologies to the market. At his venture capital office in Palo Alto, he said, two deans from Stanford sit on the advisory board and invest in the fund. "When they see something we ought to look at," he said, "they grab us by the ear and pull us there. There's a culture at Stanford that goes beyond just trying to license technology. At a lot of universities the deans don't think it's a part of their job, but I think it is."

Mr. Goldin agreed with Katharine Ku that "money isn't the main issue. Getting technology transferred is multidisciplinary, and too many universities depend on the technology transfer office. When I was in the government and traveled to the major research universities, I found a significant weakness in the process in the tech transfer shop, especially in state universities, which was the feeling that 'we can't fail.' There are too many restrictions on those individuals. They frustrate the innovators to the point where they don't want to deal with them." Mr. Goldin said he agreed with a 501(c)3 strategy that removes tech transfer from the oversight of the state so that it can be dealt with on its commercial merits. The objective of tech transfer, he said, should not be to maximize the money earned back from patents, but to maximize the value of the companies. Dr. Proenza added that the University of Akron, and "virtually all of the successful ones," were 501(c)3 organizations.

DAY 2

Welcome and Introduction

M.R.C. Greenwood
University of Hawaii

Dr. Greenwood welcomed the participants back for the second day of the symposium and introduced Hawaii's Senator Daniel Akaka. She said that he had been an extraordinary senator for Hawaii, especially in his leadership of the Committee on Veteran's Affairs. She noted further that the Senator is the first native Hawaiian and the first person of Chinese-American descent to serve in the U.S. Senate.

Opening Remarks

The Honorable Daniel K. Akaka
United States Senate

Senator Akaka said that he had listened to the speakers "with much excitement because of what this symposium means to Hawaii and our great country." He thanked President Greenwood for her invitation and gracious introduction, and expressed his gratitude to the many that had traveled from near and far to attend the conference. "I extend my heartfelt aloha. You are part of a critical effort to foster innovation and growth here in our island state."

He said that he hoped the symposium would help "set a course to make Hawaii a global leader in technology and in pioneering discoveries." He said that Hawaii was unique and had much to offer the world. Its remote location makes it one of the best places in the world to study astronomy; it has a wide array of unique endemic plant and animal species; and it has one of the most active volcanoes in the world, erupting continually for the past 28 years. Its rich cultural heritage was also unique to the state, he said, with "new discoveries going hand in hand with traditional learning. This way of life makes us unique and gives us an advantage as we compete in a global marketplace."

He commented on the motto of the symposium, *E Kamakani Noii, The wind that seeks knowledge,* which he called "a poetic demonstration of the great potential before all of us today." He added that part of the definition of *E Kamakani Noii* is "searching for even the smallest detail." He spoke of the wind, makani, as a powerful force in traditional storytelling and in daily life in Hawaii. "And so, to all of you I say: Whooo, whoo, Kamakani, let the winds blow toward a sustainable future."

He said that Hawaii was developing pioneering research initiatives, from astronomy to clean energy to sustainable agriculture. "Hawaii is a special lab for these areas of knowledge," he said. "I am proud of the many things our talented educators, researchers, and students have accomplished here. These advancements improve our way of life, stimulate our economy, and work toward environmental sustainability."

He said that innovations in research methods and practices help to preserve limited island resources and teach people to be better stewards. "Stewardship is a matter of cultural heritage and practicability," he said. "You all know the importance of maintaining a balanced relationship with nature and the necessity of caring for the land and ocean." He further called on his fellow citizens of Hawaii to ensure bountiful natural resources for generations to come.

"We all envision a better and brighter Hawaii for *akaki*, our children," he said. "And the conversations taking place through this symposium will pave the way for that to happen." He said that the symposium was "a great way to start the year 2011," and he applauded participants for their leadership, dedication to education, and to the revitalization of the economy.

Session IV

University of Hawaii's Current Research Strengths and Security and Sustainability: Energy and Agriculture Opportunity

Moderator:
William Harris
Science Foundation Arizona

Dr. Harris offered a short summary of the previous day's session to set a tone for the second day. He began with the island welcome of aloha, noting a deep fondness for Hawaii where he had attended the second and third grades. He commented especially on the quality of the new UH Innovation Council, noting the distinguished level of experience in the members. He also said that "it's a very significant thing to have some members who are from out of the state. I think that ensures that you'll have a very hard and crisp discussion, and you'll actually be able to put some new things on the table." He agreed with the senators and the other leaders of the state that Hawaii had a potential "that may be unmatched, and it is probably a state that is in the right place at the right time."

HAWAII'S SATELLITE LAUNCH PROGRAM

Brian Taylor
School of Ocean and Earth Science and Technology
University of Hawaii at Mānoa

Dr. Taylor, dean of the School of Ocean and Earth Science and Technology, said he would introduce an innovative satellite launch program that was triggered in part by a study by the National Reconnaissance Office showing a decline in the state of the U.S. space industry. "We've gone in the last decade from putting more than two-thirds of the satellites in a given year into space to less than one-third. And if you take away the military from that number, it's much worse."

Lowering the High Cost of Getting to Space

Much of this decrease, Dr. Taylor said, was caused by the high costs of getting to space from the United States, while other countries innovate to find cheaper ways. One of the ways to change the economics of space access "is to make things smaller and cheaper." The cost of developing a big satellite, he said, is about a billion dollars; even small satellites, including launch, cost about $140 million. The good news, he said, is that technology is allowing the miniaturization of technology, particularly the computing aspects. "Small satellites are going to be more and more capable," he said.

One exciting development, he said, is the development of new, space-friendly technologies such as the CubeSat, 10 cm on a side, which was mentioned the previous day by Vice Admiral Oliver of the Naval Postgraduate School. The National Reconnaissance Office, Boeing, and the Air Force are investing in this new technology, and the space office of DoD, NASA's Ames Research Center, and NASA's Office of the Chief Technologist were also promoting small satellite development.

In traditional development, Dr. Taylor said, new technologies have to be "space validated" or proven through experimental missions before they can fly. This means that a "new" technology being launched today is actually more than five years old. But today's approach, he said, is to produce components that are modular and "pre-stage" so they can be can launched earlier, with safety-redundant "constellations of small satellites." If this can be done reliably, he said, "it will be a game changer."

A New Space Flight Laboratory

In the University of Hawaii's centennial year, 2007, the School of Ocean and Earth Sciences and Technology (SOEST) joined with the College of Engineering to create a new Hawaii Space Flight Laboratory (HSFL). The partnership, said Dr. Taylor, will collaborate on every aspect of space missions, from developing spacecraft and instrumentation to mission operations and analysis. Its mission[21] is to:

- Promote innovative engineering and science research for terrestrial and planetary space missions;
- Develop, launch, and operate small spacecraft from the Hawaiian Islands;
- Provide workforce training in space mission activities;
- Promote collaboration between other institutions interested in space exploration.

---

[21]The mission of the HSFL can be accessed at *<http://hsfl.hawaii.edu/HSFL_about.html>*.

One partner is Sandia National Laboratory, which is second only to NASA in the number of launches it has performed, many of them from the Pacific Missile Range Facility (PMRF) on Kauai, Hawaii.

The first mission of the HSFL is called LEONIDAS, the Low Earth-Orbiting Nanosatellite Integrated Defense Autonomous System. Its objective is to conduct two launches from PMRF using low-cost systems and train the workers to prepare for the launches. "Fundamentally," said Dr. Taylor, "we are learning to increase access to space and get there in a different way."

The project is part of a congressionally directed program funded through the Operationally Responsive Space Office of the DoD. The space flight laboratory is the prime contractor and has multiple partners. One of those is Vandenberg Air Force Base in California, which had donated a scout rail launcher, rebuilt by HSFL. Other partners include the Aerojet Corp., manufacturer of solid rocket motor parts, the PMRF on Kauai, White Sands Missile Range, NASA/Ames Research Center, and Sandia National Laboratory. The program will use a SPARK Launch Vehicle, a three-stage solid propellant motor stack redesigned from Sandia's Super-Strypi to reduce cost, simplify launch, and increase reliability.

The Goal of a Complete Satellite Launch System

Eventually, Dr. Taylor said, HSFL aims to provide a complete satellite system and to spin off niche companies. The first of these will be a partnership between UH and Aerojet for launch services. The future may bring small satellite development companies or others in high-tech fields. It will maintain critical support facilities at UH, such as the clean room, thermo-vacuum chamber, and vibration chamber for satellite testing and spin balance. These will be for use by both the university and small businesses in the area, as well as provide "an unprecedented educational opportunity, from kindergarten through graduate school, in all aspects of space mission operations." HSFL will partner with Kauai Community College in program management and telemetry and with Windward Community College in their education and outreach through the aerospace center. In the future, a partnership may be added with the University of Hawaii at Hilo in software and automation. In addition, the space grant program allows system-wide undergraduate and high school access through the extension program funded by NASA. The community colleges will provide the technical associate degrees and the four-year colleges the baccalaureate and graduate degrees.

An additional partnership is being formed between UH and Aerojet Corp. "We're planning a 501(c)3 limited liability corporation with many benefits for each side. Aerojet will increase revenues by selling more rockets. It also hopes to set up a skunkworks for R&D in Hawaii. UH not only will gain workforce training, but will be able to fund its own science and engineering mission. The company has told us it wants to lower its costs by decreasing their overhead," he said. "That's a real driver. We have a price line to meet and they're prepared to help us meet it through this joint partnership. Together we'll handle risk management and hold the intellectual property."

The first LEONIDAS mission will launch a CubeSat planned to advance the readiness of a particular computer chip to be used in subsequent satellites for data compression. It is being built by UH undergraduate engineering students, most of whom are from Hawaii. The second mission, HawaiiSat-1, will conduct a thermal and visible image study of the earth. This will also be built by College of Engineering and SOEST faculty and engineering grad students, partnering with NASA-Ames. He showed a thermal hyperspectral imager (THI), which will measure thermal energy emitted from the surface of the earth. This tool can be used to monitor volcanoes, wildfires, urban heat sources, and trace gases in the atmosphere, such as the greenhouse gas methane. It can also detect groundwater discharges into coastal waters.

Dr. Taylor turned to cost-effectiveness. One scenario is a rideshare payload configuration in two shapes that can carry 24 CubeSats or a combination of 1, 3, 6, or 12 CubeSats, together with other small satellites. For a single CubeSat, the price of getting into space is only about $50,000. For the maximum payload, the cost is estimated at $12 million, "a fraction of the cost of any other way into space today." He said that 80 universities in 44 states were building small satellites, but they did not have a convenient way to get them into space at reasonable cost. "So they're sitting on shelves. We want to liberate that potential and get them to space."

Changing the Launch Game

Another game-changer, he said, could be the use of constellations of small satellites, which allow for more efficient packaging of payload. He showed a planned series of altimetry missions for the decade 2010-2019, including two ways of carrying out a major mission. One way was a billion-dollar satellite called SWATH. A different technique, using three small satellites to accomplish the same functions, he said, would cost less than $100 million.

Researchers at the UH are interested in accurate rendering of ocean color, which is necessary to monitor the health of coral reefs. Current and planned spacecraft do not do this adequately, but the hyperspectral imager planned at the UH is designed for just this function. Similarly, Dr. Taylor plans to include methane, a potent greenhouse gas, in the UH observing program. Methane, which has begun to rise again after remaining constant through the 1990s, is currently not monitored from space.

Dr. Taylor summarized by saying that the innovative satellite launch program of the UH and its partners is poised to make an original contribution that is low in cost, low in risk, and capable of rapid response (less than one week). The involvement of the university in the program promises not only a new economic driver for Hawaii but also a focus for developing the high-tech workforce of tomorrow. The program's first launch is planned for 2012.

ASTRONOMY IN HAWAII

Robert McLaren
Institute for Astronomy
University of Hawaii

Dr. McLaren, associate director of the Institute for Astronomy,[22] said that he would summarize the development of modern astronomy in Hawaii and describe how that development contributes to innovation and technology transfer. He began his story in the early 1960s, when space science was expanding rapidly during the Apollo program. In 1960, a tsunami devastated the area around Hilo, and the head of the local Chamber of Commerce, Mitsuo Akiyama, was looking for projects to spur the economic recovery effort. He championed the idea of placing a major astronomical observatory on nearby Mauna Kea and sent letters around the world to observatory directors. He got only one response, from Gerard Kuiper of the University of Arizona, one of the world's leading planetary astronomers.

In 1963 Kuiper, then director of Lunar and Planetary Studies at the University of Arizona, was on the summit of Haleakala, on Maui, where the UH was installing the Mees Solar Observatory. The site was a good one, but occasionally enshrouded by clouds. Kuiper and his assistant, Alika Herring, would look from Maui across the Alenuihaha channel at another mountain peak on the Big Island, about 65 miles away. At nearly 14,000 feet altitude, it was 4,000 feet higher than Haleakala and above the clouds. Perhaps remembering Akiyama's letter, Kuiper decided to take a closer look. He chartered a plane and flew over the summit, which was broad enough for multiple observatories. Now truly excited, he went to see Governor John A. Burns and convinced him to put a Jeep trail from the mid-level of Mauna Kea to the summit. This was finished in a few months, and by the summer of 1964 Kuiper had established a site testing station. Dr. McLaren showed a photo of Dr. Kuiper and Governor Burns at the summit. "You can see that astronomy activity in Hawaii had strong gubernatorial support from the beginning," he said, noting the continued support of the current governor, Neil Abercrombie.

The View from Mauna Kea

Once Kuiper and Herring had confirmed that Mauna Kea was the best site for astronomy they had ever seen, they assumed that the establishment of research programs there would be led by mainland universities, and particularly the

---

[22]The Institute for Astronomy (IfA) was founded at the University of Hawaii in 1967 to manage Haleakala and Mauna Kea Observatories and to carry out its own program of fundamental research into the stars, planets, and galaxies. It has a total staff of more than 300, including about 55 faculty members. *<http://www.ifa.hawaii.edu/ifa/about_ifa.shtml>*.

University of Arizona. However, after many preliminary steps and despite many obstacles, the University of Hawaii, which had no nighttime astronomy program at the time, was chosen to establish the first research telescope on Mauna Kea, a 2.2-meter facility that is still in use today.

Today there are 13 telescope facilities on Mauna Kea, Dr. McLaren said, representing a capital investment by 11 countries of over $1 billion. Development of the program has been supported by both the governor and legislature of Hawaii. The state paid for the infrastructure, including the road, and set aside about 13,000 acres (later reduced to 11,000) for a science reserve, where astronomy could develop with a buffer against other activities. The congressional delegation has lent strong support, as have federal agencies, especially NASA and the National Science Foundation. The county of Hawaii has also helped in many ways, especially by passing an ordinance to limit light pollution.

Since the Institute for Astronomy was created by the university to guide the astronomical aspects of the development, it has entered into partnerships with 10 other organizations on the U.S. mainland and other countries. The partner organizations build the telescopes and pay for the operations, while the university maintains the site, helps initiate the programs, and in return shares observing time.

Protecting the Mountain

For the first 35 years, the Institute of Astronomy was responsible for virtually all aspects of the development on Mauna Kea. Today, while it still provides guidance on the scientific programs, the University of Hawaii at Hilo and its Office of Mauna Kea Management manage public access, community relations, and protection of the environment and culture. "This has been a major change," said Dr. McLaren. "It's still something of a work in progress, but in my opinion a great success, and a model for how other entities could approach challenges like this. Large programs that start as a seed activity naturally outgrow the ability of the initial group to manage all the aspects and address community concerns."

Dr. McLaren displayed a photograph of the observatories, with Mauna Loa in the background. Keck 1 and Keck 2, each with 10-meter segmented mirrors, are the largest telescopes of their type in the world.[23] Other observatories include Subaru Telescope, Japan's eight-meter facility; the Caltech Submillimeter Observatory and the James Clerk Maxwell Telescope; Gemini North, which, like its southern cousin in Chile, is run by a consortium of countries; and others.

The next large project being planned is the Thirty-Meter Telescope, or TMT, another ambitious consortium project. Current members include Caltech, the University of California, and a group of Canadian universities, while "interested and future partners" include Japan, China, and possibly India. This is a billion-dollar project, currently in the permitting phase, with ground-breaking anticipated in the

---

[23]Mirrors larger than about 8 meters in diameter cannot be made from single blocks of glass. Larger mirrors are made of multiple small segments precisely fitted together.

next 12 months. A notable feature of the TMT is that it is not quite on the summit but on the northwest plateau where it is less visible—a concession to local beliefs in the sanctity of the site.

Activities on nearby Haleakala, like those of Mauna Kea, are planned and regulated jointly by the Institute of Astronomy and by a local body, in this case the UH Maui College. The original Mees Solar Observatory has been joined by the Air Force's Maui Space Surveillance Site, and the mountain, also like Mauna Kea, is awaiting a new and larger facility, the Advanced Technology Solar Telescope (ATST), a $300 million project of the U.S. National Solar Observatory. Its 4-meter mirror will be the largest of its kind. The reason for its large size is that major questions of solar astronomy depend on analysis of the magnetic field and the sun's surface at scales of tens of kilometers. The ATST is also in the permitting stage.

Tracking Dangerous Asteroids

Another project on Haleakala, the Pan STARRS 1, is unusual in being explicitly a project of the Institute for Astronomy. Pan STARRS has a mirror of modest size—about 2 meters—but it also has the largest-capacity camera in existence, a charge-coupled device of 1.4 billion pixels. This camera, built at the UH Institute, has such a wide angle it can survey the entire sky about 20 times per year, mainly looking for objects that change or move. It has many practical applications, especially in tracking asteroids whose orbits might bring them close to the Earth, as well as new supernovas and variable stars.

Astronomy activities at the UH has also led Hawaii toward a leadership position in cyber-infrastructure. Because astronomy is highly data-intensive, he said, the Institute had taken the lead in equipping the state with the wide bandwidth required by the observatories. In 1996, the observatories partnered with Hawaiian Telephone, contributing $2 million to help install fiber optic cable across the saddle of the Big Island. With the help of the National Science Foundation, DoD, and others, UH extended the cable further so that high bandwidth communication is now available to anyone.

The UH has also developed strong local instrumentation capacity, building instruments for numerous telescopes and satellites. These activities have led to spinoff activities, notably the company formed by former faculty member Gerry Lupino, called GL Scientific, which is based in Honolulu. Another innovator, Doug Toomey, an engineer at the Institute for 25 years, runs a small business in Hilo building astronomical equipment. "Having these activities in Hawaii," concluded Dr. McLaren, "brings benefits beyond the actual technology. In addition to technology transfer to the marketplace, we transfer experience and enthusiasm to school kids, neighbors, and other people who see what is possible in this whole fascinating field. Hawaii's kids get to see world-class science and technology in action every day."

DATA ANALYTICS: A PROPOSAL

The Honorable Daniel S. Goldin
Intellisis Corporation
and 9th NASA Administrator (Ret.)

Mr. Goldin began by saying that he wanted to propose a strategy for Hawaii that he had contemplated for some time, and which was reinforced by the report of President Greenwood's Innovation Council and the work of Dr. Taylor, Dr. McLaren, and other UH faculty. This strategy was to develop regional expertise in what he called "smart software." He said that he used the term smart software to signify the varieties of data analytics, network analysis, artificial intelligence, and machine learning that are growing rapidly.

He noted that several speakers had already referred to the nation's "data-hungry" culture and the information being generated in quantities soon to overwhelm our ability to store and, most importantly, use it effectively. He said that a "crossover point" occurred in 2006, when data creation equaled the world's total data storage capacity. At present, he said, data-intensive programs, such as the Large Hadron Collider at CERN, are producing more data than current techniques can process. "Many places are just throwing it onto the floor," he said, "because they can't analyze it. There's a crying need for some smart system to be able to pick out essential pieces of information."

And it's going to get worse, Mr. Goldin said, as more wireless platforms add to this overflow of information. Today there is almost one cell phone per person in the world, and most of those are used just for phone calls and text messages. By 2013, he said, there will be 2 billion smart phones, which use much more data, and these will be followed by tablets like the iPad, which use still more data. Finally, he said, in 5 to 10 years "the scientific vision of 'smart dust' will come to reality." Smart dust refers to minute wireless particles that serve as sensors in health and many other applications, transmitting their data via the Internet. "Instead of having millions of sensors," he said, "we'll have trillions of them that will assist us in doing unbelievable things. States or regions or clusters that start thinking in these terms now will be ready to deal with the future."

Economic Opportunities from the Data Glut

Mr. Goldin said, the economic opportunities in this area are large. He said that current information technology (IT) spending in the United States was about $400 billion and was projected to grow to $800 billion by 2017. "Of the increase," he said, "almost all of it is from smart software."

Hawaii, he said, was not yet positioned to take advantage of this trend. The state was currently last in the United States in broadband capacity, as measured by the percentage of the population with access to more than 5 MB in broadband capacity. What tools does Hawaii need to capture this business 10 years from now? he asked.

The first is access to broadband technology, which was very low. "Leadership of this state must deal with that," he said.

Second, Mr. Goldin said, was a highly skilled indigenous work force. "Your university is capable of producing this work force if the state would bring its support," he said, and move beyond the "internecine warfare" that was blocking real educational progress. "It's not a question of who wins what," he said. Programs should include not only students from community colleges through the postdoctoral level, but also the children from K through 12 "so they have some sense of the future."

Third, he said, the state must develop an "innovation friendly ecosystem." On a personal note, he said he had tried several times to start businesses in Hawaii, and he had "flunked each time." On each occasion, he said, he had been prepared to make an investment but "could not deal with the lack of a friendly ecosystem." Finally, he said, there must be access to investment capital located "here and not on the mainland."

Why Smart Software for Hawaii?

Mr. Goldin then listed a series of reasons why smart software was an appropriate business opportunity for Hawaii:

1. **Location**. Hawaii, located centrally in the fast-growing Pacific region, had the potential to develop partnerships, clients, and customers both in the United States and prosperous eastern Asia. "Your population is well connected to Asia," he said, "and they want to do business with you."

2. **Research and academic strengths**. He said that Hawaii is an acknowledged leader in Asian cancer research, astronomy, and other fields. "With smart software, you can both do better science and bring new business to Hawaii."

3. **DoD supercomputing center**. This facility brings another local opportunity for expertise in smart software. The U.S. military facilities on Hawaii have many data needs and can serve as a communication node for the rest of world. "Why bring all the information in and just relay it to the mainland for processing? he asked. "Why not process it and send it by narrow band back to the mainland? Take the data and turn it into knowledge yourselves."

4. **Smart software bypasses supply chain problems**. One handicap of Hawaii's remote location is the difficulty of interacting with supply chains in manufacturing. Smart software firms require few parts or materials.

Mr. Goldin offered an illustration of the economic opportunities available to a region with talent and resources in smart software. Many businesses, he said, can benefit from new, fast-growing techniques of social networking analysis. He described some research done by IBM for Bharti Airtel, a large Indian communications firm, which needed an edge to protect its business from competitors. IBM, tapping into sociological research data, designed a product of social networking software that would allow the company to identify certain types of customers. For example, the IBM software showed that the most influential people in a given region were those who made long outgoing calls and short incoming calls; these would presumably be the best potential Airtel customers. "Think of the value of protecting some of the existing tourism business in Hawaii by using smart software," he said. He noted that similar social networking techniques had led to the discovery of Saddam Hussein by demonstrating that before and after terrorist attacks, the drivers of certain cars would get more phone calls.

Seizing an Opportunity

Mr. Goldin suggested that within the next decade, Hawaii could be in a position to offer just such services and to build a broadly competitive software industry. In order to do so, he said, the state needs to accomplish the following:

1. **Adopt the 2008 Broadband Plan for Hawaii**. He praised this document, which predicted large paybacks from broadband improvement. For example, it showed that a 7 percent increase in broadband adoption would have a $578 million economic impact. And yet the state, stymied by disagreements between local jurisdictions, had not yet implemented this plan.[24] "At present," he said, "Hawaii has just two more slots for fiber optic cable landings. After that, no more will arrive unless you change the way you do business."

2. **Build an indigenous workforce of the best and brightest**. Such a work force can be trained in the UH system, he said. Just 1.4 percent of the state's work force was employed in the mathematics and computer science sector, he said, compared with the national average of 2.5 percent. "I would set a goal of roughly doubling this work force in 10 years. In 20 years you ought to be up to 10 percent and you can become dominant in broadband."

3. **Build an innovation friendly ecosystem**. This is essential in order to nurture new businesses and attract existing ones.

Mr. Goldin argued that the state had everything to gain from such a strategy, and that a determined state leadership could break the existing logjam preventing its adoption. In his view, the state had the potential to build a smart software industry capable of not only servicing local business sectors, but also of moving

---

[24]The final report of the broadband task force is available at <http://www.hbtf.org/>.

into the global market place. "You can start as soon as you get the critical investment funds, which I don't think is a problem with a state of your resources. Hawaii has everything it needs to become a leader in the information technology of the next decade. The critical components are leadership, speed, and the right partnership. And the time to start is now."

HAWAII: A MODEL FOR CLEAN ENERGY INNOVATION

Maurice Kaya
Hawaii Renewable Energy Development Venture (HREDV)

Mr. Kaya, project director of HREDV, said he would talk about ways to prepare Hawaii for the energy future by making it a model for innovation. "We've heard a lot about the innovation ecosystem," he said. "I think there is no better place in the world to gain from an energy ecosystem than Hawaii. This is partly because our needs are so great; we import everything, and for over 20 years, we haven't moved the needle at all. But I remain optimistic. We have to do something right away, and if we do not, we are going to be at the tail end of suffering."

He said that the state, to its credit, had put together a "wonderful" Hawaii Clean Energy Initiative (HCEI) to move toward the "unheard of" goal of 70 percent clean energy by 2030. While he applauded the target itself, he warned against "just importing the technology and the dollars without gaining the benefits from those investments. I would like to suggest that along with this transformation we so urgently need and seek in the energy markets, we see tremendous opportunity for innovation to be derived from those investments." One analysis by the state has placed the needed investment for this transformation at around $18 to $20 billion over this period, he said. "We would be remiss if we did not take advantage of those investments to create our own innovations and benefit from them."

Some Benefits of Clean Energy Technology

Mr. Kaya listed some of the lasting benefits to be expected from a successful clean technology sector in Hawaii:

- Local, high-quality, high-paying jobs;
- Opportunities for Hawaii's youth;
- Industry leadership and opportunities to export innovative technology;
- Import substitution and energy security;
- Diversified income streams for agricultural land;
- Reduced greenhouse gas emissions.

He said that the road to this goal would not be easy. "As we move into these unprecedented levels of transformation and into clean energy markets, we will face problems and challenges," he said. "We have a lot of wind—*makani*—and a lot of sun. But they are not here all the time. We have to find a better way to utilize these intermittent sources of energy." Other challenges, he said, included integrating photovoltaics, deploying efficiency at scale, integrating electric vehicles with the grid, developing new biofuels, and developing new energy sources for agriculture.

A broader challenge, Mr. Kaya said, was to make costs manageable and predictable at the same time. On the brighter side, some innovative companies in Hawaii have already begun this process, and in some cases are well under way: he mentioned Pacific Biodiesel, Sopogy, Oceanit, Concentris Systems, Referentia, HNUTechnologies, Hoku Scientific, Makai Ocean Engineering, and Honolulu Seawater Air Conditioning, LLC.

A Catalyst for Clean Energy

His own organization, the Hawaii Renewable Energy Development Venture, was created in 2008 as a catalyst for the local clean energy industry. "We recognized," Mr. Kaya said, "that investors who support these types of companies need to be comfortable with their investments." The HREDV is designed to support this need through three strategies:

1. **Competitively awarded funding**. Competitions help accelerate the commercialization process and support local and mainland companies investing in commercialization activities in Hawaii.
2. **Training and other capacity building**. For new entrepreneurs, managing federal funds is not an easy process, he said. HREDV provides training for young firms needing assistance.
3. **Strategic partnerships**. The HREDV seeks to be a catalyst and believes that facilitating partnerships among industry players and coordinating with multiple levels of government are primary responsibilities.

Mr. Kaya said that the focus of HREDV is on technologies that are nearly ready for the commercial marketplace. He showed a chart of the Technology Readiness Levels, from basic technology research through system operations, emphasizing that the pipeline needed to be constantly refreshed. "This is where the university can have such a significant role—in making sure we have this continual pipeline of technologies that allows more and more efforts to reach commercialization and bear fruit."

So far, HREDV had supported five projects in Round 1, totaling $2.1 million of federal funding that was matched by $1.4 million of private cost sharing. In selecting these companies, they had tried to listen to the barriers described by the energy community, the electric utility, the state, and state consultants. The first round focused on issues such as green transportation, from both a fuels and vehicles standpoint, and also on agriculture, a fossil fuel user linked to both import and export.

Mr. Kaya said that because his organization had little funding, it has to leverage what it has by picking "potentially great companies with great innovations," along with significant cost sharing. Another emphasis is to recognize these energy activities as parts of an integrated system.

Clean Energy Start-ups

As an example, he said that Concentris was a local company that had developed wireless mesh technology for military applications. It was attempting to use this in partnership with another local company, Oceanit, to address the daunting problem of non-metered energy consumption in military housing. They worked in a public-private partnership with one of the major housing contractors for the Navy in Hawaii, Forest City Military Communities.

Another example was Sopogy, a pioneer of Micro-Scaled Concentrating Solar Power, or MicroCSP technologies. MicroCSP uses concentrating mirrors with optics, low-cost thermal storage, sun tracking, and a simplified installation technique. They can be installed on rooftops, and a rooftop array coupled with absorption chilling was being developed for the Maui Ocean Center. The technology can also generate electricity from solar heat, providing a non-photovoltaic alternative for general commercial cooling.

Satcon is a company from the mainland east coast which was responding to a common problem of clean energy: the supply of PV-generated electricity varies with available sunlight, and the charging of electric vehicles adds grid load unpredictably; both can affect grid stability. Satcon is developing an inverter to allow efficient charging of vehicles using direct DC solar power, as well as smooth solar power for a better interface with the electricity grid to be initially demonstrated on Lanai.

Another local company, Kuehnle Agro Systems, was a product of the University of Hawaii. It is trying to address a general problem in biofuels, which is that large tracts of agricultural land are often required. Kuehnle is building a pilot bio-reactor to produce customized algae screens for companies that want to produce biofuels from micro-algae.

The last company Mr. Kaya highlighted was Better Place, a software company that is installing infrastructure for nine electric vehicle charging stations. It plans to use seven electric vehicles to provide the first demonstration of integrated vehicle-to-grid technology on Oahu, partnering with the Hawaiian Electric Company.

"These are the types of innovations we're supporting," he said. "I believe that energy is where it's going to happen in terms of making a mark for this state." To achieve this goal, he said, innovation is critical, and the success of innovators depends on access to the market. At the same time, the traditional incumbents in the energy sector will come to depend on the success of these innovators. "The energy system of the future," he said, "will allow you as a customer to have more ability to establish how you want your energy supplied and what you are willing to pay for a certain level of quality. It will be very different from the vertically integrated energy systems we have today."

The same problems faced by Hawaii, Mr. Kaya concluded, are faced by regions throughout the United States and the world. "So with our success, there's a very high prospect we can be a world leader not only as a model for energy innovation, but in the way we apply these technologies for everyone's benefit."

SUSTAINABLE AGRICULTURAL SYSTEMS: CHALLENGES AND OPPORTUNITIES

Sylvia Yuen
College of Tropical Agriculture and Human Resources
University of Hawaii at Mānoa

Dr. Yuen, dean of the College of Tropical Agriculture and Human Resources, said that she would talk about the broad context of sustainable agricultural systems. "When the word 'sustainable' is used," she said, "very often people think about a particular practice. Let me be clear this morning that when we talk about sustainable we will not be referring to a specific practice, like organic farming, or an end point, but rather a broad systemic strategy." She said that sustainable agricultural systems should satisfy human needs, enhance environmental quality and protect the natural resource base, promote economic health, and enhance the quality of life. "So it's really a multi-dimensional process which considers all of these goals from the outset rather than limiting itself to one goal at a time."

She said that the most remarkable feature of American agriculture is that "it has been amazingly successful." The U.S. population, she said, has increased more than four times since 1900. "But despite the growth in our population, there are fewer farmers at work today, and they are producing more food and fiber for the domestic and export markets. And we're doing all of this on basically the same acreage of land we used a century ago."

Even so, Dr. Yuen noted, great challenges remain. "It's estimated that one in seven people around the world are still malnourished. And the situation is going to be worse. We're at about 7 billion people globally right now, and by the year

2045, it's estimated that there will be 9 billion people. How are we going to feed all those people? she asked. "The simple answer is, do what you've done in the past. You've been successful; do it again. The solution is not that simple, however, because there are many challenges that we haven't faced in the past." She listed these challenges:

- Increase in income
- Competition for land, water, and energy
- Impact on the environment
- Effects of climate change
- Cost of production

Impacts of the Challenges

Today, Dr. Yuen said, more countries are experiencing rising incomes. "And when people have more money, they eat differently. They consume more meat, more fish, more dairy products, and more processed foods. This changes the way the food chain is constructed. This is because producers will follow the money and the demand. So the kinds of foods and the way we grow them will be different." She noted that producing one pound of meat takes three pounds of grain.

There is progressively more competition for land, water, and energy, both from outside of agriculture, as in urbanization, and from other agricultural activities. The increased competition for non-food uses of crops, such as the demand for biofuels, also means that lands are being taken out of food production.

The practice of modern agriculture also has serious impacts on the environment, Dr. Yuen said. Part of the success of American agriculture is owing to its strong and singular focus on food production and on reducing costs. "But this focus, we now know, resulted in agricultural practices that produced some unintended consequences," she pointed out. The same fertilizers and pesticides that boost food production also have a detrimental effect on ground water, rivers, and soils—not just with respect to human health, but also in terms of the environment. "And these are things we can't allow," she said, "as we move forward to a sustainable agricultural system."

The Challenges of Agriculture in Hawaii

A particular difficulty for Hawaii, Dean Yuen noted, is the high cost of land. "The economic return from building houses and shopping centers is much greater than growing food." She said that in 2006, when sugar was declining as a crop, some lands no longer tilled were slated for housing and priced at $44,000 an acre. This, she said, contrasted with the $1,000 to $5,000 an acre generally paid in the rest of the nation for agricultural lands. "You can see that this makes it quite prohibitive to acquire land to be used for agriculture. If you are a large landowner and you look at the economic returns, it is to your advantage to keep those lands fallow or limit them to short-term leases, which is what most farmers in Hawaii have. If you're a farmer, you have little incentive to put in long-term investments, even if they could help you become profitable later. That means no energy efficiencies, no planting crops that might take longer than a few years to mature."

Production costs are also high, Dr. Yuen said, averaging $434 per acre, in contrast with $261 per acre for the nation as a whole. "Our farmers tell us that they will farm if they can make money. But we have to create the conditions that make farming and food production economically feasible."

Dr. Yuen listed some additional challenges for agriculture in Hawaii:

• Hawaii is particularly vulnerable to food supply disruptions, both natural and manmade.

• Some 85 percent of the food consumed in Hawaii is imported. "Ironically, instead of becoming more independent, despite our efforts, we are becoming more dependent in some areas on imports. For example, in 1984, 100 percent of the milk consumed in Hawaii was produced locally. Today it's 30 percent." Hawaii's farms are mostly small and diverse, which reduces efficiency. About 64 percent are farms of 10 acres or less, versus 11 percent on the mainland.

• The risk of invasive species is unusually high because of the high level of importation. The ships, airplanes, and tourists bring disease-bearing organisms, harmful insects, weeds, and other pests that can negatively affect agricultural production, the natural environment, and the economy in general. The imported Mediterranean fruit fly, which lays its eggs on more than 400 fruits and vegetables, has reduced the yields of many crops, including papaya, guava, and mango. The coffee berry borer, introduced recently, which lays its eggs in the coffee berry, is becoming widespread and reduces production.

Some of Hawaii's Advantages

In the face of these challenges, she said, Hawaii does have many advantages as an agricultural region. It has 11 of the 12 soil types found in the world, and 10 of the 14 climatic zones. "This means that we have a rich living laboratory for research that can be of value to many parts of the world. Inhabiting Pacific islands, we have geographic commonality with two-thirds of the world's population, enjoying the same climatic conditions and growing many of the same kinds of crops on farms of similar size. So much of what we do here," she said, "can be used in other parts of the world." She noted that our research, capacity-building expertise, and cultural competence can enhance food security and economic stability in other countries, which can ultimately contribute to world peace.

Dr. Yuen said that if Hawaii really wants to help feed the world and to grow in global importance, "we have to expand and enrich our research initiatives." This includes studies of sustainable practices, genomics and genetics;

decision-support systems in which we simulate conditions; integrated pest management; and a transformative approach in which large, interdisciplinary teams use their collective efforts on site-specific areas.

She also talked about work based on an ancient Hawaiian concept in agriculture called "ahupua'a." "The practice was to look at the whole tract of land, from the mountain to the ocean, and pay attention to the interplay of all the factors—soil, climatic conditions, vegetation, wildlife—and how each part interacts with the others and contributes to the whole. We don't necessarily do that in modern times. Our unit of analysis has been the farm or a single-use geographic area. When you focus that way, the interplay of all of these variables, including the human dimension, gets lost. So perhaps it would be useful to bring back some proven concepts from yesterday to combine with our best science today."

Dr. Yuen concluded with the thought that "this is a very exciting period to be an agriculturalist, to be engaged in the research and the very hard work of putting what we learn in the lab into practice. Where else can you use your talents and your intellect and all of your skills to feed the world, protect the environment, and improve people's quality of life?"

DISCUSSION

Teena Rasmussen, a member of the UH Board of Regents, proposed the flower farm she has run jointly with her husband for 32 years as one model for agricultural development. The 50-acre farm, Paradise Flower Farm, is located in the Kula Agriculture Park on Maui and inhabited by several dozen employee farmers. At the end of the 1970s the county had determined that the park was an appropriate place for agriculture and had put in water lines and roads, leaving farmers to develop the land with the benefit of 50-year leases. The Rasmussens cleared the lots, purchased water meters, and brought in power. The farm has been a success, she said, not only providing fresh flowers and lei in Hawaii and on the mainland, but also offering an enlightened working environment for its employees and farmers. As the farmers grew older, the Rasmussens changed the ordinance so leases could be assigned and sold. This means that if farmers build a building or greenhouse on the land, they can recoup their investment in the building. Now the farms are turning over, and new owners are coming in to continue the farming. The land has been designated for farming in perpetuity. She encouraged the state and counties to promote more such developments on what are termed Important Agricultural Lands (IALs) to preserve Hawaii's agricultural capacity and way of life.

Dr. Yuen affirmed the value of this agricultural model. She said that the model addresses two key challenges faced by farmers: affordable land and adequate water. These and other challenges are more difficult than the work done in the laboratory, she said, because they're being resolved in the arena of the real world of politics, competing values and interests, and trying to convince people to use new science. She said that she did see in agriculture a readiness for change. In the last legislative session, the Food & Energy Security Act was passed, which places a tax on every barrel of petroleum that enters the state. Part of that revenue is allocated to energy independence and food self-sufficiency and safety. The expense of providing water for farming was an equally pressing issue, she said, and should be addressed with public-private funds.

Session V

Medical Opportunities in Hawaii

Moderator:
Virginia Hinshaw
University of Hawaii at Mānoa

Dr. Hinshaw, chancellor of the University of Hawaii, introduced the final session by noting the importance of health care in Hawaii, whose population "lives longer than any other in the United States." She said that health care was an appropriate local emphasis, with the "health sciences increasing in strength at UH Mānoa." She cited the words of Thomas Jefferson, who said, "Public health is what we as a society do to ensure conditions in which people can be healthy. The care of human life and happiness, and not their destruction, is the first and only objective of good government. Without health, there is no happiness."

She emphasized the collaborative work of the health sciences, in which Hawaii sought to integrate its particular strengths of basic sciences, nursing, medicine, social work, cancer research, and public health. "We envision this as a system-wide endeavor," she said, "including pharmacy and allied health efforts across the university. For much of my career, I worked internationally on influenza viruses, and I learned early that countries need to work together to keep our local and global neighbors healthy."

CLINICAL TRIALS IN HAWAII

Art Ushijima
Queen's Health Systems/The Queen's Medical Center

Mr. Ushijima, president and chief executive officer of the Queen's Health Systems and the Queen's Medical Center, noted that even after 37 years' experience in hospital administration, he was newly excited by several innovative programs in Hawaii's health care. One was a new partnership between the medical school, which does not have a hospital, and the hospitals in the community. A second was the hospitals' growing experience with clinical trials. He noted that Hawaii's uniquely diverse population made it an ideal location for clinical trials, and that true local expertise in the medical care of Asian Pacific populations was likely to draw patients not only from Hawaii but also from throughout the Pacific Rim region.

He said that the Queen's hospital was founded in 1859 by Queen Emma and King Kamehameha IV and has been in its current location since July 1860. Today it is a tertiary care teaching hospital affiliated with the John A. Burns School of Medicine of the University of Hawaii. The hospital admits over 22,000 inpatient cases a year and 350,000 outpatient visits. It employs about 3,500 people, including 1,100 physicians.

Today, he said, a focus of Queen's is to build a strategic framework around innovation. This was not easy to do in a large hospital, he noted, because so much of its activity is driven by tradition and a broad array of regulatory requirements, but he was seeing exciting improvements. Some of these improvements were in process, such as using clinical protocols to standardize medical care. Others came from program development, such as wider application of minimally invasive surgery, including a robotics surgical program. A third area is research and development, he said, primarily in the area of clinical trials. He saw three primary areas of value in the clinical trials work: (1) It brings novel therapies to patients, (2) it feeds an innovation framework by supporting researchers developing new therapies, and (3) it brings opportunities for commercialization of new drugs, devices, and applications.

Mr. Ushijima described three successful clinical trial initiatives that have been developed.

Three Successful Initiatives

Queen's Center for Biomedical Research

This center, he said, is an outgrowth of recruiting Reinhold Penner, M.D., Ph.D., from the Max Planck Institute in Germany 13 years ago. Dr. Penner's wife and professional collaborator, Andrea Fleig, Ph.D., earned her doctorate at the UH and wanted to return, and Queen's worked with the university to recruit them. It built a laboratory and provided seed capital for his research program, while simultaneously strengthening the laboratory capacity on campus for him and for other researchers. It also increased the National Institutes of Health (NIH) indirect funding rate from 19 percent to about 75 percent today, developing a new funding model by sequestering indirect funds to support the investigators for special equipment and for times when they were needed. The funds were not used for operations, but to seed and support research.

Much of Dr. Penner's research had addressed cell signaling, signal trans-duction, and ion channels. At the same time, he had experimented with the drug clofazamine, known for treating Hansen's disease, and was beginning to apply it to innovative treatment of autoimmune diseases, such as psoriasis, rheumatoid arthritis, multiple sclerosis, Type 1 diabetes, and Crohn's disease. Such practical applications can extend the value of existing drugs and provide new treatments for patients.

MRI Research Center

Mr. Ushijima said that Queens's has also worked closely with the medical school in developing novel solutions for known problems. One example grew out of the desire of a researcher, Linda Chang, M.D., to move from Brookhaven National Laboratory in New York State to Hawaii in the early 2000s. The focus of her research was the effects of methamphetamine on the brain and HIV on the brain. When she received a grant for a three-tesla MRI from the Office of National Drug Control Policy, Queen's committed $2 million to building out new space and providing some ongoing laboratory support. Dr. Chang brought her team and more than $20 million in grants to the university. Her husband and professional collaborator, Thomas Ernst, Ph.D., is a researcher in digital imaging who is developing a technique to address a negative feature of MRIs. An MRI may take 45 to 60 minutes to complete, and if the patient moves, it may have to be repeated, which is costly, time consuming, and disruptive for the patient. Dr. Ernst and collaborators have developed a technology to provide real-time correction to an MRI image that might otherwise be blurred by motion. Mr. Ushijima said that he estimated that reducing the number of MRIs in the United States alone could save over $1 billion in repeat MRIs.

PET Imaging Center

A third initiative, Mr. Ushijima said, grew out of an agreement established in 1998 with Hamamatsu Photonics. The company, internationally known for photo-multiplier tubes used in satellites imaging and astronomy observatories, wanted to site a PET scanner prototype in the United States. They proposed a joint venture with Queen's to build a $5 million cyclotron facility to manufacture the radio-pharmaceuticals for PET imaging while Hamamatsu Photonics provided the pro-totype PET scanner. The timing was good, because Medicare in 1998 recognized the value of PET and began to reimburse for procedures. Recently collaborators have developed a tracer that offers a solution to a detection problem with prostate cancer. The common tracer for prostate cancer, fluorodeoxyglucose, accumulates in the bladder and masks the prostate. The Queen's researchers developed a tracer process for a choline compound that does not accumulate in the bladder, but goes to the prostate, where it is visible in a PET scan. The choline technique, able to give better staging, avoid unnecessary biopsies, and track response to chemo-therapy, is now going through Food and Drug Administration (FDA) approval, and is also being tested on breast cancer, hepatoma, and brain cancers.

Recruiting Senior Researchers

Results from each of these centers feed directly into clinical trials, Mr. Ushijima said, and each of the projects has brought valuable lessons to Queen's. First, he said, it is important to recruit established and funded researchers. "Having the expertise is critical." Second, the institution must provide an adequate infrastructure to support the researchers. Third, leading researchers must be able to pursue collaborations and strategic partnerships all over the world. And fourth, the institution must develop "path to market" expertise. This fourth ability, he said, had not yet been developed, but was a current focus at Queen's. "One thing that Barry Weinman has imbedded in us," he concluded, "is ROI, return on investment. Whatever you do, you have to include a plan for ROI and a way to sustain it."

Mr. Ushijima added in conclusion an essential fifth ingredient: the willingness to take some risk. "In all these initiatives," he said, "our board at Queen's was willing to assume a significant amount of risk. This is what has allowed these facilities and projects to happen, and we've had significant success as a result. But we still have a lot more to do."

UNIVERSITY OF HAWAII MEDICAL INITIATIVES

Jerris Hedges
John A. Burns School of Medicine
University of Hawaii at Mānoa

Dr. Hedges, dean of the School of Medicine, began by placing the medical school in a developmental context. He said that the school had had "pockets of excellence that predated my arrival" and provided a "nucleus for our matrix." The current strategy, he said, was to "start bringing clusters of knowledge generation together, and linking them with the community at large."

In the year 2015, he said, the school of medicine will celebrate its 50th anniversary as a medical school. It trains not only doctors, he said, but also public health officers, medical technologists, speech pathologists, and basic scientists. "We have probably the most diverse student body population and faculty of any medical school in not only the United States, but in the world." Of the current medical school class, 64 were chosen from 1,800 applications; 90 percent of the students are residents of Hawaii. He said that the school's excellent Kaka'ako Campus had opened in 2005, made possible through the generosity of taxpayers; the cost of bonds issued to permit the construction came from the state's share of the tobacco settlement agreement. On the same campus, construction is also under way for a new University of Hawaii Cancer Center.

More Community Partnerships

The current strategy for the medical school, he said, was to bring education and research together and to build a complete biotechnology enterprise, in partnership with the community health care institutions. He said that an overarching goal was to move further into community partnerships. An example was the new biosafety research cluster within the basic science building that allows research on significant infectious diseases, such as dengue fever, avian flu, West Nile virus, and HIV/AIDS.

Dr. Hedges praised the neuro-imaging group directed by Dr. Linda Chang and described by the previous speaker, as well as the Institute for Biogenesis Research, developed by Dr. Ryuzo Yanagimachi. This lab had developed some of the original techniques of in vitro fertilization, as well as techniques of transgenesis, the placing of genes from one family of organisms into another, including the first demonstration of moving genes from one organism into the DNA of another. Some original cloning work had also come out of the Institute for Biogenesis Research, he said, helping us understand how birth defects evolve during cell multiplication.

In a related program, the school participates in the National Children's Study in which 1,000 Oahu children are to be carefully monitored from the pregnancy of the mother through their lifetime. At the other end of the age spectrum, he said, the school's Department of Geriatric Medicine is nationally recognized, partly for its nationally ranked educational and research programs tailored to Hawaii's diverse communities. The Kuakini Medical Center, for example, is known for its studies of the mechanisms and genetic basis of healthy aging. Another partner was the Department of Native Hawaiian Health, in which investigators have been addressing the role of social stressors on the health of indigenous peoples.

Dr. Hedges closed with a "plug for my colleague Michele Carbone," who was studying a disease caused by the mineral erionite in Turkey, where it is commonly used in the construction of churches and other structures. Dr. Carbone and colleagues found that erionite, a mineral similar to asbestos, has a strong association with the lung and peritoneal disease mesothelioma in some families, but virtually no association in others. This had led to an international collaboration to discover the genetic basis of this difference.

The Melding of Ethnicities and Cultures

The medical school has strong NIH research programs, he said. These programs are well suited to study the melding of ethnicities and cultural groups and have allowed researchers to ask questions beyond simply what is the molecular basis of illness and to look into social and ethnic factors. "One Clinical Translational Research award allows us to begin on multiple levels to start unraveling the issue of nature versus nurture," he said. The school focuses on six key health areas to identify causes of health disparities amongst (ethnic, cultural, and geographic) groups: cardiovascular, respiratory, cancer, nutrition and metabolic, perinatal and growth, and aging/neurocognitive health. "We look at the community at large," he said, "trying to identify factors that promote better health in our population. We also provide resources to young investigators to help prepare the next generation of investigators."

The medical school, like other academic institutions, is also trying to translate the results of its research into useful form. In the health sciences, this has to do with understanding how knowledge at the basic level is taken to the patient's bedside, and from there to the community at large. One goal is to better understand existing disparities between groups. Is the difference simply due to access to early diagnostic studies and therapies? "Why do population subsets differ? We have a great opportunity to study that issue," Dr. Hedges concluded. "In each of the six key health areas we are focused on documented disparities within Hawaii, and we have expertise in each of those areas. We will start to unravel those factors and hope our multicultural, multiethnic population can serve as markers or bellwethers for health in the country as a whole."

One way to translate innovations to the community, he said, is through the embryonic thrust of personalized medicine, ethnically and culturally tailored. Another way is through new business models that are health focused, use alternative delivery systems, and minimize redundancy in care delivery. "We need to know why some treatments work for some individuals and not for others," he said.

Changing the Medical Model

The larger challenge, Dr. Hedges stated in closing, was to make fundamental transformations in the practice and delivery of health care. "Some 17 percent of GNP is now devoted to health care," he said. "That has to decrease. We have to deliver medical care in a more healthy prospective manner. We have to introduce social impact factors and personal involvement in one's own health care. This means a move away from an insurance-driven transactional model, an end to our process of waiting until someone becomes desperately ill and then rescuing them from that state.

"We also have to change our definition of medicine from what the physician is doing to what the health care team is doing. It's time for the medical school, nurses, and social workers to work closely with hospital partners to make that change. In our lifetimes, we've seen a lot of industries come and go, and I think our health care delivery system is going to change too. We need to move fast to be ready for it."

ADVANCING INNOVATION AND CONVERGENCE
IN CANCER RESEARCH

Jerry S. H. Lee
Center for Strategic Scientific Initiatives
National Cancer Institute
National Institutes of Health

Dr. Lee, deputy director of the National Cancer Institute (NCI) Center for Strategic Scientific Initiatives, noted that cancer is known to be not one but many diseases, and that many are "relatively manageable" when the disease is detected early and localized. But the thing that connects all of these many cancers is what happens when the disease spreads, or metastasizes, and contributes to more than 90 percent of cancer deaths.

Approximately half a million Americans died from cancer in 2010, and about 1.5 million will be diagnosed this year. "What's startling to me is that, while we've made progress in other diseases since 1950, we've not done such a great job with cancer, which challenges us to think more innovatively about the disease. By 2020, the experts tell me that mortality will rise to about 10 million per year worldwide and incidence to about 16 million worldwide." Citing recent publications by the World Health Organization (WHO), in some regions, he said, incidence and mortality have increased by 50 percent since 2002.

However, he said, there has not been a more exciting time for research given the advancements in both information exchange via the internet and the many new technologies that are bringing deeper understandings of the human genome and variations that may relate to cancer at a unprecedented pace. For example, less than 10 years after celebrating the complete sequencing of the human genome, in 2010, the NIH announced the 1000 Genomes Project to establish the most detailed catalogue of human genetic variation among different ethnic groups. Dr. Lee said while considerable new data is being generated, the statistics he cited earlier clearly demonstrate that transforming the data into new knowledge to generate value to the patient has been more challenging than anticipated. One reason, he said, is the rising costs the pharmaceutical industry faces to develop each new drug—costs that are unsustainable by current models.

How to Make a Large Difference Fast

In 2003, when the NCI started the Center for Strategic Scientific Initiatives (CSSI), the primary objective was to help figure out a way to break out of this unsustainable high-cost paradigm. The goal was to identify steps that could make a large difference right away. The Center found that the pace of progress was slowed by "turning the crank thinking." When the community of researchers was asked directly what should be done to break the paradigm, he said, "We were a little shocked at the answers." The survey of researchers said that faster progress depended on better standards, protocols, real-time public release of data, multidisciplinary teams and thinking, and, more importantly, individual team members who themselves were trained in more than one discipline. "You can imagine that these are very tough tasks to deliver. But we figured that if it would give us the potential to transform cancer drug discovery and diagnostics, we should do it."

Dr. Lee stepped back to put the NCI in context. The NCI was created as part of the National Cancer Act, signed into law in 1971 by President Nixon. From a recent report generated for the National Cancer Advisory Board, currently about 50 percent of cancer research is conducted by private industry, and a quarter supported by NCI. "We have heard a lot today about partnerships, and this is one area where they are strong."

He said that the NCI Center for Strategic Scientific Initiatives had launched approximately seven programs since 2003, all of which "took key inputs from the research community" and recognized "that some of the things we're doing will, in the next 5-10 years, yield real benefit, and some are already doing so." Dr. Lee then gave an example from a nation-wide nanotechnology initiative, led by Dr. Piotr Grodzinski, that just the previous week launched its third clinical trial which uses not only a nano-based platform to stage the patients but also will give the patients the appropriate nano-enabled therapeutic.

Dr. Lee said that the Center found after eight years of work that a good idea does not equal innovation. However, a good idea paired with some unique implementation will significantly increase the potential for innovation. This pairing had already led to some unanticipated innovations.

Launching a Cancer Genome Atlas

"So our very first step," Dr. Lee said, "was to go back to the drawing board and ask: Where did we miss some potential opportunities?" Cancer has always been thought of as a disease of the genes, but everyone had their own way to characterize these changes in small sample sets. He said we took the community's first recommendation of standardization and decided to do a systemic identification of all genomic changes at a large-scale by multiple teams, repeat it for all cancers, make the data publicly available in real time, and test whether the data could be assembled in a format akin to a chemical engineering steam table.[25] Efforts to accomplish this started in 2004, and in 2007 the Center launched The Cancer Genome Atlas (TCGA) program.[26] It focused on three pilot diseases—brain, lung, and ovarian cancer—to see if the concept would be helpful. What the group first encountered when they tried to do this for hundreds of samples was that there

---

[25]A steam table lists the properties of steam at varying temperatures and pressures.
[26]<http://www.genome.gov/17516564>.

was a national shortage of highly annotated biospecimens for cancer research. "This was an unanticipated innovation on our part," he said, that led to an ongoing effort, led by Dr. Carolyn Compton, to build a public resource called the cancer human biobank (caHUB) to provides such highly annotated biospecimens to the research community.[27]

Returning to TCGA, Dr. Lee noted that despite the cautions of many naysayers, state-of-the-art genomic characterization of more than 400 cases of glioblastoma (GBM) by several expert teams working together, followed by a subsequent unprecedented multi-disciplinary analysis of the collected data and existing available clinical data about the diseased tissue, revealed several unanticipated discoveries. This was published in a landmark paper in a 2008 *Nature* article, not only because of the wealth of data, but also due to the fact that the manuscript had only one author, The Cancer Genome Atlas Network. And since the data were made publically available in real-time, prior to the publication, several scientists not supported by the program made their own unanticipated scientific discoveries and that led to new hypotheses. For example, clinicians had used the GBM reference cancer genome and "now are able to separate responders from non-responders." In 2010, researchers identified a new subset of GBM that occurs in younger patients, and a technique to better predict outcomes. "So we're really excited by the way this has enabled the entire research enterprise to innovate."

He said that at the same time when the Center was preparing for the TCGA program, the community had asked whether a similar atlas could be created for cancer proteins, which are encoded by genes being examined in the genome atlas. Unfortunately, as assessed by the National Academies, technologies used to detect protein changes were not yet mature enough as compared to those used by TCGA to examine genomic alterations. As such, the Center launched a program in 2006 to get proteomic technologies ready for clinical utility. Led by Dr. Henry Rodriguez, the program announced, in 2009, the first successful multi-site assessment of proteomics tools in the United States, including a public data resource for those who want to look at the raw data set. The program also began working with the FDA to develop a "mock FDA 510(k) document"[28] to educate new developers on how to clinically prepare proteomic technologies for FDA approval to be used in clinics. Also, the Center established an antibody characterization laboratory that gives out highly characterized, high-quality antibodies and reagents at nominal costs. He said that CSSI would soon launch the second phase of this program, pairing genomic abnormalities found in TCGA with their associated changes in proteins.

---

[27] *<http://biospecimens.cancer.gov/cahub/default.asp>*.

[28] FDA Section 510(k), or Premarket Notification, requires medical device manufacturers to submit premarket notifications if they intend to introduce a device into commercial distribution for the first time.

Enlisting the Perspective of Hard Science

Dr. Lee noted to the audience that all of the programs he mentioned thus far seemed only to be contributing to the already extraordinary amount of data being generated by the research community. He said this was something the Center recognized back in 2008 and knew that a new perspective was needed in order to sort and integrate these large and complex datasets. CSSI asked a group of 300 physical scientists, engineers, mathematicians, and other "hard scientists" that were used to combining vast information to help think through whether basic principles and laws could be developed for cancer, like they were done for fields such as physics and engineering. Over a span of a year they asked "people who had never thought about the disease to think about it with us." They responded with some valuable insights: for example that the cancer community was thinking of signaling pathways in two dimensions, when in fact they occur in three dimensions. "Proteins move around," he said. "They don't just sit next to each other." He said that the physical scientists were fascinated by metastasis, and how rare, random, and inefficient the phenomenon is, and yet so deleterious. After several additional workshops, the Center launched in 2009 the Physical Sciences-Oncology Centers (PS-OC) program that brought together life and physical scientists led by Dr. Larry Nagahara. "We gave them a charge not to just give us better science, but paradigm shifting science, and to build transdisciplinary teams to achieve this. We gave them some tools we've never before given to NCI grantees—such as the ability for individual PS-OCs to give out internal and external pilot funding to support what the grantees felt were important new innovations."

Dr. Lee closed by inviting members of the workshop to reach out to the 12 awarded PS-OCs and contribute their own expertise and form collaborations. "I think you have some interesting perspectives that these teams would love to hear about. Also, it helps that we gave them the mechanism to exchange personnel, and it's probably not going to be a very hard sell to have one of these professors send a post-doc or graduate student to Hawaii to collaborate with your existing infrastructure." He thanked the organizers for the opportunity to describe "what we see as the potential for the future of personalized health medicine."

BIOMEDICAL INNOVATION WITH GLOBAL IMPACT IN HAWAII

Hank Wuh
Skai Ventures
Cellular Bioengineering, Inc.

Dr. Wuh, who identified himself as "a proud product of the University of Hawaii," said that he had such a "fabulous education" there that he managed to get into medical school after only two years. After doing his residency at the UH School of Medicine, he moved to orthopedic surgery at Stanford University where he also started his first company. He had moved back to Hawaii seven years ago to take care of his father and to run his company Skai Ventures.

Why Start-ups are "Really Hard"

He began by acknowledging that start-up companies "are really hard" and gave a brief personal reminiscence about the experience. "The truth is," he said, "there's nothing sexy or glamorous about start-ups. Resources are tight, it's difficult to raise money, the risks are great, and outcomes are unpredictable. Adventurous entrepreneurs do it because they believe in a concept, an idea, and a dream. In a fair fight, the big guys will always win. They have the resources, the infrastructure, and the network. Start-up companies and entrepreneurs are always the underdog. So you have to be nimble and creative. At Skai, we are the champion of the underdog and like to give them a chance so one day their dream might grow up to become a giant. As you all know, it does happen. It doesn't happen all the time, but if you don't try, you will never succeed."

Dr. Wuh said that visitors to the Skai office would see three words on the wall: Invent, Disrupt, Inspire. "Invent is obvious," he said, "but the world disrupt is there for a simple reason. We like to focus on working on the sort of technology that is truly transformational. Rather than things that are a little faster, smaller, or cheaper, we'd like to focus on technology that can fundamentally change the future. At the end of the day, it's no more work or time or energy to do that versus doing something incremental. And finally, to inspire: That refers to our interns. This past summer we had 15 interns, all children from Hawaii. The youngest was a junior in high school and the oldest had just finished college. They spent the summer with us to learn about technology and entrepreneurship. It's a fantastic group, because they're young, smart, and energetic. Most importantly, they're completely fearless, because they don't understand what limitations are yet. They don't have grown-ups' fears about how things can be impossible."

Champion for the Underdog

The Skai model is a very simple one, said Dr. Wuh. It scours the world for great ideas from really smart people, typically at major universities and national labs. "Our job is to help them, because we are the champion for the underdog. The trick is we bring these innovations back home to Hawaii, where we grow them into big enterprises."

Skai works with many partners, he said. In addition to private equity funding, Skai has put about $35 million into their portfolio of seven companies. They also work with government partners who advance the technology and help mitigate some of the R&D risk. To illustrate the excitement of new technologies and the breadth of opportunities for a small venture firm, Dr. Wuh summarized the work of several Skai companies:

- **Eyegenix**. This company grew out of a chance meeting between Dr. Wuh and a Japanese ophthalmologist. The eye doctor mentioned that Asians were often reluctant to donate organs, mostly because of cultural or religious beliefs. Worldwide, corneal transplants are the number one human transplant procedure and perhaps the most successful. But some 10 million people in the world are blind from cornea disease, and the world supply of donor cornea can only address 100,000 cases. So 99 percent of those needing corneal transplants, he said, will not have a chance to see. In the United States, there are almost enough donors, but the problem is severe in Asia. Skai decided to scour the world for a technology that could be used for an artificial cornea. He found one at the University of Ottawa, under development for 13 years, and got an exclusive license for the technology. At Sweden's Karolinska Institute, Skai found the surgeon who had done the pre-clinical studies with this technology, and about 2.5 years ago took the first group of patients to him. The first procedure was done on a 75-year-old woman, blind from a genetic disorder, and when her cornea was replaced with the Eyegenix material, she regained her vision completely, as did a chef, a school bus driver, and the rest of the 10 patients. The data was published in 2010, approvals are being sought, and a manufacturing plant is being built on the site of an old Dole cannery in Hawaii. "It's a big deal for us here in Hawaii," he said. "We can manufacture enough supply for the world's cornea transplantation in a facility less than 5,000 square feet in size." He said his company is also building an international treatment center where eye surgeons have invested as partners, and entered into an agreement with Visionsure, the largest consortium of eye banks, to provide donor corneas. The eye clinic is planned to become the foundation for a broader program of medical tourism, taking advantage of the state's experience in dealing with visitors and extra capacity of hotels during certain seasons.
- **CBI Polymers**. In the process of developing artificial corneas, Skai built expertise in materials science—specifically in a binding property between the polymer and the surrounding environment. When the Air Force put out a request for proposals for a polymer gel that could bind radioactive particles, Skai responded. "We knew nothing about airplanes, and nothing about radioactive particles," he said, "but we knew a lot about a polymer's specific binding properties." From that idea has evolved a series of products to remove many kinds of toxic materials, including nuclear contamination, to restore building surfaces, and to perform many other tasks. "There is opportunity to do something significant," he said, "even though we're in Hawaii." CBI Polymers now has 50 customers around the globe.

Dr. Wuh closed by saying that there are many opportunities to do something "really significant here in Hawaii. It's about creating a way that we can allow entrepreneurs to flourish in our community. These are the young people that we really want to help and inspire. Innovation is by nature difficult, and start-up companies are challenging. We have to help them in every way we can."

DISCUSSION

Dr. Greenwood asked what the university could do to help create and bring more innovative companies to Hawaii. Dr. Wuh said there are two components. One is the culture or belief system. "In order for people to take risks, they have to understand that it's okay to take risks. Entrepreneurship means taking chances on early-stage ideas and concepts. But in order to implement that, you have to have resources. We don't want to create great companies and smart entrepreneurs only to have them leave the islands. They need resources here to build and thrive and be successful right here in Hawaii. There are many ways to achieve that, but it takes an entire community to believe that is an important part of the future."

Dr. Hedges acknowledged the "wonderful example set by Dr. Wuh in coupling knowledge with business." At the medical school and affiliated programs, he said, "we have been able to generate a lot of good ideas with commercial potential, but we need individuals with both motivation and know-how to couple with our business community." He said that this process was a challenge for the health professions and that colleagues from the business school had been "extremely active" in putting programs together for entrepreneurs. "We also have to change what we're expecting of our faculty, whose thinking is still limited to traditional publishing, grants, and promotion, rather than how they can work with the business community and be part of the economy."

Elmer Ka'ai of the UH said that "this was probably one of the best events I've attended in the last 10 years," and asked for advice on ways to pull the various communities together. He noted a Hawaiian proverb that says, "This banana tree is exceedingly strong only because everybody got together to produce this single fruit." He noted an inability of the medical community "to really move into Hawaiian communities without scaring them away." How might we produce an environment, he asked, that brings in the Hawaiian people, who have so many health needs?

A Department of Native Hawaiian Health

Dr. Hedges commented that "we are blessed with having a Department of Native Hawaiian Health, which, in large part, was generated by a grant from the Queen's Medical Center. "We [at the School of Medicine] followed the lead from Queen's in building programs around the involvement and needs of native Hawaiians. We have native Hawaiian leaders who have ties with the community in a variety of areas. The leadership in the department has developed networks with health centers and practitioners across all the islands to participate in an outreach program, and it's through this network that we hope our research program will be community engaged. We have such a diverse culture here in Hawaii that it's impossible to fully understand all the cultural issues that exist, but we've put a lot of effort into having our students culturally sensitized so they can appreciate that they have limitations and can reach out for appropriate consultation. I think any efforts from the school are driven more by what the community has asked of us, and one of our challenges is to share our knowledge so that what we know can be more useful for their needs."

Dr. Wessner asked Dr. Hedges, in regard to the traditional orientation of the medical school faculty, what might be done immediately to change incentives toward innovative behavior, such as giving tenure credit for work with start-ups or patents. For Dr. Lee, he asked what kinds of activities at the medical school might attract new investment by the NCI.

New Incentives for Faculty

Dr. Hedges replied that, in fact, within the past month, the process of promotion and tenure had been reviewed, and the institution had decided to shift its reward system toward teamwork and partnerships of various kinds. "We want to reinforce that the landscape has changed and we're expecting a different level of performance on the part of our faculty." A related incentive, he said, was to reward contributions to collaborative (U54) grants and accomplishments in the private sector, such as the generation of patents and work with start-up companies.

Time to Invest in Data Analysis—For Medicine

Dr. Lee responded to Dr. Wessner's question by referring to a point made earlier about informatics and support of computer infrastructure. "I think that would be a very wise investment down the road because many of these postdocs and graduate students who are working with medical and clinical issues are looking at the data and beginning to use it for patient care." Data analysis, he said, has the potential to find molecular signatures that can be "worth more than any new drug." It is time to invest in that activity, he said, because the need is growing more quickly than the analytical or storage ability. NIH is running out of storage space, he said, and The Cancer Genome Atlas is having to send discs of data to other centers for processing. "I think there's some real power to be investing both in the manpower and in the informatics and IT infrastructure."

Dr. Hedges added that a previous dean was a "champion of problem-based learning" who had changed the curriculum. While this was an extremely positive step, he said, it meant that much student teaching was based on the ability to access information online and to synthesize that information in solving problems. "Getting essential information to make a clinical decision will be more and more challenging," he said. "We need to enjoin our colleagues from computer science and mathematics to help us do a better job in synthesis and integration of that information."

Mr. Goldin agreed that data challenges are imminent for Hawaii. He added that he had been asked to estimate the minimum data capacities for the state, and had answered that 100 megabits of capacity should be provided for every home in Hawaii, and 1-10 gigabits for scientific research and business research. "These are absolutely mandatory thresholds," he said, "if you want to have a high-tech industry that's stable and grows." In addition, he said, the computer science capacity and supportive technologies at UH are not sufficient. He suggested a war chest of at least $10 million to $20 million to hire the best people and upgrade the infrastructure.

Session VI

Roundtable—Next Steps for Hawaii

Moderator:
M.R.C. Greenwood
University of Hawaii

Participants:
The Honorable Mazie Hirono, United States House of Representatives
The Honorable Colleen Hanabusa, United States House of Representatives
Peter Ho, Bank of Hawaii and APEC 2011 Hawaii Host Committee
Richard Rosenblum, Hawaiian Electric Company
Donald Straney, University of Hawaii at Hilo
Charles Wessner, The National Academies

CONGRESSWOMAN MAZIE HIRONO

Congresswoman Hirono referred to Hank Wuh as one of the "great innovators and disrupters in our community," and emphasized that "we need to identify them and support what they're doing." She recalled the governor's first task force on science and technology, convened 15 years ago, which reported on the importance of recognizing the University of Hawaii as an economic engine and the importance of commercializing the research there. "Here we are," she said, "basically talking about the same thing. We do know what we need to do, it's just the doing of it we haven't mastered." She said that she had read the Innovation Council's report with approval and said that "nothing could be more important than the education component." She noted that one of its recommendations was to encourage the entrepreneurial spirit at the university, "but we all know it needs to begin long before college. This is why it's important for us to support robotics in our public schools, for example, to be sure we have white boards or smart boards in our schools. I visited classes where the kids are really engaged by using technology, and we already know that. I know that everybody here has great ideas, but the task is truly to stay the course. We have been talking about this for a long time in this state."

CONGRESSWOMAN COLLEEN HANABUSA

Congresswoman Hanabusa said she appreciated the wonderful opportunity "to actually see something start." She was referring to President Greenwood's first State of the University speech in February 2010, from which she cited three main points. One was the need for money for infrastructure, "and the university did receive funding for that." A second point was that Hawaii has to invest in the workforce that the state will need to develop areas of particular strength, including cancer research, alternative sources of energy, vulcanology, astronomy, and information technology. A third point is the need to look toward the future in a focused way. "We all have to leave here feeling that we have defined where we want to be and how we can get there. We're not going to be the best in everything. Let's decide what we can realistically be the best in, and let's funnel our resources to achieve that."

PETER HO

Mr. Ho, who said he would speak in his capacity as president of APEC, cited the management professor Peter Drucker in observing that "in business, there's innovation and there's marketing, and everything else is just a detail." "As a finance person myself," he said, "I take a little offense to that, but I think there's a lot of sense to it as well." Mr. Ho said that APEC could potentially mean a great deal for Hawaii's business community, supported in powerful ways by innovation and marketing. With the APEC meeting in November 2011 in Honolulu, "really what we're talking about is a world-class exposure opportunity for our community." Honolulu would host President Obama and other leaders from the 20 largest Asia-Pacific economies, who will be accompanied by their ministers. They would be joined by upwards of 1,000 CEOs and senior executives and some 2,000 members of the global media, along with "general members of the global community," the representatives of think tanks, and other organizations that follow global trends.

Joining the Wider Asian-Pacific Community

"With that opportunity for us to host people who don't often head into our neck of the woods," he said, "come broader opportunities for innovation as well as for marketing." In terms of innovation, Mr. Ho said, an obvious one is the opportunity to "put a stake in the heart" of the problem of how to evolve from just a "sun, sand, and surf marketplace in the visitor industry" to a place for serious meetings, conventions, and other business settings for the visitor industry. Another opportunity, he said, is to use the event to "help drive our own community in its thinking from primarily domestic" to more actively international, as part of the wider Asia-Pacific community. Finally, he said, the state can make good use of the Hawaii Host Committee, which he had chaired for the past year. This truly representative committee, he said, represents "a truly different approach to doing things than we've had in the past. People will come for business conventions as they recognize how serious we are about innovation."

In addition, he said, would come obvious marketing opportunities, including clean energy technology, the economic power of the UH research programs, and the growing expertise in the health sciences. "These are strengths that people don't always attribute to Hawaii, but in fact we do have those resources, and APEC is simply a great opportunity to expose them to the world."

RICHARD ROSENBLUM

Mr. Rosenblum, a nuclear engineer, said he had been recruited just over two years previously to be president of Hawaiian Electric after 32 years in the utility business at Southern California Edison. In late 2008, the state of Hawaii, along with the Department of Defense and several other constituencies, made and signed an agreement to "break Hawaii's addiction to oil," he said. The goal was to reduce the consumption of fossil fuels in Hawaii by 70 percent in a little over 20 years. "To the best of my knowledge, there is no country, state, or institution that has ever been that brave because at the same time they are saying we don't know how to do this, but we're going to find a way."

This bold step, he said, was economically essential. In 2008, the domestic product of Hawaii was nearly $60 billion, of which just over $8 billion was spent on fossil fuels. "Think of an economy with almost a 15 percent hole in the bottom, draining the economic vitality. It is not survivable." The goal of the state, he said, was to plug that hole, using indigenous resources to the extent possible. "This agreement wasn't a choice, it was a necessity."

About one-third of fossil fuel use goes to long-distance transportation, he said. Another third goes to short-distance cars and trucks, and the final third to electricity. Electricity is the most straightforward to address, he said, because "there's one monolith in that sector. As I said, I was trained as an engineer, and when the National Academy of Sciences says that global warming is real, stop arguing.[29] Global warming is real, and we're not going to change it unless the whole society decides to end our dependence on fossil fuels and go to clean energy sources." He added that this presents a strategic opportunity for Hawaii— not just to reinvigorate the economy, but also to create jobs.

---

[29]The National Academies has published more than 15 publications about the science and policy implications of global warming and climate change. See, for example, National Research Council, *Understanding Climate Change Feedbacks*, Washington, D.C.: The National Academies Press, 2003.

Developing and Exporting Clean-energy Knowledge

A clean energy mandate also creates intellectual capital, Mr. Rosenblum said. "The bad side of the initiative is that we don't know how to do it. The good side is, we're going to figure it out. And when we figure it out, we'll have knowledge other people will need. The reasons for moving Hawaii off fossil fuels are the same reasons the rest of the world will eventually have to move off fossil fuels. And when they do, the people who did it first and developed the knowledge have a unique opportunity to capitalize on that. So while Hawaii is not going to be a manufacturing complex, even for windmills, we can be an exporter of knowledge and drive our economy with things we've learned."

DONALD STRANEY

Chancellor Straney, of the University of Hawaii at Hilo, said that he, too, had come to Hawaii only recently, but he already felt a difference. He said it would have been unlikely to have a discussion "as exciting and promising as this one in the state I came from, California. There is a sense of future and a sense of promise here, and in education, that's what we need to keep things moving forward."

He added that he wished for a way to translate and communicate some of the symposium to every third-grader in Hawaii, the level "where people or students make their commitment to technical and scientific futures and get the fundamental excitement that can shape their lives in a productive way. It would be interesting to take these topics down into the schools, as Congresswoman Hirono mentioned, and use them as a hook to draw students forward."

Learning to Use Knowledge in "Real Life"

Chancellor Straney said he agreed with President Greenwood's desire to make innovation and entrepreneurship part of the University of Hawaii curriculum. UH Hilo was then in the process of writing its vision statement, he said, and one component under discussion was a requirement that all students take an internship or other practical experience in their discipline.

"We're very good at teaching people theoretical foundations," he said, "but in my experience, it's when you go out and actually use knowledge that you begin to understand how it works in real life. We need to help students understand what to do with a good idea when they have one.

In regard to technology transfer, he noted that other universities where he had worked managed this function through an independent foundation. Aside from managing patents and licensing, he said, there were many things a university can do to help small business, such as providing access to the instrumentation on campuses and develop partnerships with people who don't have access to cutting-edge equipment. "This can make a huge difference to a start-up company or someone moving into a new technology area." He said that at his previous institution in Los Angeles, "We simply made it known that we would rent time on the scanning electron microscope we had, and that drew a lot of interest from the small businesses in our vicinity. They wanted not just to use that piece of equipment, but also to talk with us about how we could help meet their needs through expertise."

Finally, he said, the challenge of the tensions that sometimes arise between the various cultures of Hawaii might be turned to advantage. He said that as a biologist, he recognized that the easiest way to generate innovation in the natural world is to isolate an organism and let it respond to local conditions. Hawaii is certainly isolated, he said, and provides contexts and experiences not available on the mainland or the rest of the world. "To what extent can we find ways to make what is unique about Hawaii, particularly the cultural aspects, motivate and inform what we do?" He referred to Dr. Wuh's "really trenchant example" of recognizing that in a culture's resistance to organ transplantation lay an opportunity for innovation that could change things not only for that culture but for people around the world. "I think it would be helpful for us to reflect fairly deeply on how to take advantage of the tensions between the cultures we have, and to view them positively rather than as constraints that needs to be overcome."

CHARLES WESSNER

Dr. Wessner praised the very high level of the presentations at the symposium, as well as the state's positive attitude toward its innovation challenges. "One of the things you've talked about a lot here is your diversity and your ethnic roots. But what I hear is actually America ringing here. And as Americans, we are interested in solving problems, not just talking about them."

He summarized this positive attitude by saying that, in contrast to many other states, "you get it. You get the need for innovation. And the thing I did not know, and I don't think our committee knew well before today, is that you have the platforms for innovation. You have superb opportunities in energy, health, astronomy, and other areas, and you can see how to use them."

At the same time, said Dr. Wessner, the state will have to work hard to generate the revenue it needs to drive innovation in those promising areas. He urged the state's leaders to start with senior government officials who had come to Honolulu from Washington, D.C., to participate in this symposium. This group, which includes Ginger Lew of the National Economic Council, Barry Johnson of the Economic Development Administration, Jerry Lee of the National Cancer Institute, and Roger Kilmer of the Manufacturing Extension Partnership program, "represents an outreach from the federal government that understands what you are trying to do."

Another pressing need he addressed was equipping the university with the resources it needs. "You already recognize the role of the university, which is to serve as an engine of economic growth. You already have effective presidential leadership in place at the University of Hawaii. Now you need to give that president the authority needed to capitalize on the resources and opportunities the university provides. In some cases, this may mean removing regulations that were designed to fit the problems of the last century, thus removing impediments to the opportunities of the future."

Removing Barriers to Innovation

One of the messages he would propose, Dr. Wessner said, is to "clear out the junk." The state is held back by mindsets, regulations, and other barriers to innovation and new business formation, he said. He referred to Mr. Goldin's comments about energy and about bringing more broadband capacity to Hawaii. "You want to make it easy to come here, not difficult. You want to make it attractive to stay and to start a business. With a million people, you won't be able to do this entirely alone, so build that environment to welcome skilled people, capital, and ideas, and to retain the graduates of this excellent university."

He also urged local leaders to develop new funding mechanisms, especially an innovation fund for the state and a seed fund for the university. He referred to Mr. Goldin's suggestion of a "war chest" that could allow the leadership in the university "to build up a department, to meet a need, and to adjust that need." He supported the idea of expanding or directing the use of the barrel tax for some of this. "I know that nobody likes a tax, or anything that sounds like a tax. But OPEC taxes us regularly, and one response is to tax ourselves to work out from under OPEC's tax."

Dr. Wessner closed by complimenting the state, university, and business leadership for "getting it," and urged still more focus, speed, and critical mass. He praised the leadership of Governor Abercrombie and urged the state legislature to "step up and help provide the programs, the space, and the reforms that are necessary. I don't think you can stay where you've been, and the need is to move from there with some speed."

DISCUSSION

Dr. Greenwood recalled the remarks of Dr. Walshok about "how you motivate people to come together and focus on certain initiatives, or, as we prefer to say here, how you get everybody in the canoe paddling in the same direction." She said that lessons from San Diego "are ones I hope we can use, because we will need the support group, and we will need to forge partnerships with the university, the business community, and the citizens. Partnerships have to be built with real commitment to common objectives, and not just a cordial association." She also reminded her audience that San Diego was a "barren field" when Dr. Walshok and her colleagues began building programs, and "tourism was a mainstay. It started from two or three or four committed people and companies, drew a lot of help from the localities, and then things began to blossom."

She said that in a small state like Hawaii, people have to work together and to remove barriers. "It has to move beyond the one-industry dominance," she said. "A big thing comes and hangs on for 10 or 15 years and then is gone."

Building a New Model for Hawaii

To build a new model, Dr. Greenwood suggested that the first step is to "be sure we've got what we need to be competitive, and connected to the rest of the world." This included the speed of Internet connections, access to those connections, and wise use of broadband resources.

Second, she said, "let's take these areas where we already have prominence and dominate them. Let's not just be good, let's be the leader."

Third, she urged her colleagues to "find those areas where we're not yet dominant but have some very special opportunities." From the university's point of view, she proposed the health sciences and the research-based businesses that can flow from that field. "We do have some strength in the health sciences," she said, "and lately it's been growing. But compared to other major research universities we can probably improve by over 50 percent and maybe as much as 100 percent by attracting investment both at the federal and the state levels, and overcoming some of the situations that make it hard to do business here."

Dr. Greenwood closed by saying that the group had learned a great deal in the past two days, and it was now time to act. "It's not a time to discuss what we could do, it's now time to discuss how to do what we know we need to do," she said. "If we don't do it, this small but wonderful state of Hawaii is not going to be on the cutting edge in two decades. If we do act, we may be an example that everyone else admires, and I'd sure like to be part of that story."

Congresswoman Hirono added that Hawaii is "not in this alone, because as a country we have recognized the need to innovate and to become much more successful in competing with the rest of the world." She said that during her last four years in Congress, she had supported an innovation agenda that provided loans, grants, and stimulus funding for alternative energy and research. "I know this administration is also very intent on working with us so that our whole country can move forward."

Broadband as "Mission Critical"

As the symposium concluded, several participants returned to the need to improve broadband access, data storage and processing abilities, as well as IT and software skills. Dr. Greenwood assured one speaker that the issue had been a priority since her arrival several years ago, and had only grown in significance.

"This is not just another priority where some progress would be nice," she said. "This is 'mission critical' for the state. We have to recognize that we're not going to be able to expand our astronomy program, develop smart software, or attract the high-tech businesses we want unless we have these tools."

Dr. Greenwood asked for comment from David Lassner of the UH, who chaired the broadband task force, and he provided the "good news" that Recovery Act funding was allowing the delivery of fiber optic cable to every public school, college, and library in Hawaii, which would provide one major piece of the state's broadband needs.

The final speaker, Chuck Gee, a member of the UH Board of Regents, applauded this particular good news. He praised symposium participants for their action-oriented discussions. "I didn't know what I expected to find after spending two days here," he said. "We're a state that's very good at planning, and some of our plans go back a long way, and they're still sitting on someone's shelf. But now I am reminded by these speakers how really good a university we have, and I see this university becoming a catalyst for making things happen. That is the new wrinkle. The kinds of fields we have heard about are compatible with who we want to be. Tourism is a clean industry, and so are clean energy, space research, health sciences, and digital media. Speaking for myself and many of my fellow regents who have sat through this symposium these past two days, you have our total support, and thank you to our visitors for bringing a new sense of excitement as to where we can go from here."

III

APPENDIXES

Appendix A

Agenda

E Kamakani Noiʻi
Wind that seeks knowledge

A Symposium Organized by
The U.S. National Academy of Sciences
and
The University of Hawaii

January 13-14, 2011

Kauaʻi Ballroom
Sheraton Waikiki
2255 Kalakaua Avenue
Waikiki, Hawaii

DAY 1—JANUARY 13, 2011

9:00 AM   **Welcome**
Oli by *Kealiʻioluʻolu Gora and Lilikala Kameʻeleihiwa*
Howard Karr, Chairman, University of Hawaii Board of Regents
M.R.C. Greenwood, President, University of Hawaii
Mary Good, Dean of Engineering, University of Arkansas at Little Rock

9:15 AM   **Opening Remarks**
The Honorable Daniel K. Inouye, United States Senate

9:30 AM   **Presentation of the Hawaii Innovation Council Report**
M.R.C. Greenwood, President, University of Hawaii

10:15 AM **Coffee Break**

10:30 AM   **Session I: The Global Challenge and the Opportunity for Hawaii**
Moderator: Tyrone Taylor, President, Capitol Advisors on Technology

The Innovation Imperative and Global Practices
Charles Wessner, Director, Technology, Innovation and
Entrepreneurship, The National Academies

State and Regional Economic Context
Carl Bonham, Executive Driector of UHERO and Associate
Professor of Economics, University of Hawaii at Mānoa

Focusing Federal Resources: The Obama Administration
Innovation Initiatives
Ginger Lew, Senior Counselor, White House National Economic
Council

12:00 PM   **Lunch**

12:30 PM   **Luncheon Address**
The Honorable Neil Abercrombie, Governor, State of Hawaii

1:15 PM   **Session II: Leveraging Federal Programs and Investments for**
Hawaii
Moderator: The Honorable Brian Schatz, Lieutenant Governor,
State of Hawaii

The Manufacturing Extension Partnership: The Network Effect
Roger Kilmer, Director, Manufacturing Extension Partnership
Program, NIST

DoD Strategic Technology Capability Thrusts: Opportunities to
Fuel Hawaii's Innovation Economy
Starnes Walker, Chief Engineering and Technical Director,
University of Hawaii

The Military and Higher Education
Vice Admiral Daniel Oliver, USN (Ret.), President, Naval
Postgraduate School

Infrastructure for the 21st Century Economy: The Role of EDA
Barry Johnson, Senior Advisor and Director of Strategic Initiatives,
Economic Development Administration

3:00 PM   **Coffee Break**

3:15 PM   **Session III: Small Business, Universities, and Regional Growth**
            *Moderator: Keiki-Pua Dancil, President and CEO, Hawaii Science
            and Technology Institute*

            40 Years of Experience with Technology Licensing
            *Katherine Ku, Director of the Office of Technology Licensing,
            Stanford University*

            **Universities and Economic Development: Lessons from the
            "New" Akron University**
            Louis Proenza, President, Akron University

            Converting University Research into Start-Up Companies
            Barry Weinman, Co-Founder, Allegis Capital LLC

            Improving Industry Partnerships
            *Mary Walshok, Associate Vice Chancellor of Public Programs and
            Dean of Extended Studies, UC San Diego*

5:00 PM   **Closing Remarks**
            M.R.C. Greenwood, President, University of Hawaii

DAY 2 — JANUARY 14, 2011

9:00 AM   **Welcome and Introduction**
            M.R.C. Greenwood, President, University of Hawaii

9:05 AM   **Opening Remarks**
            The Honorable Daniel K. Akaka, United States Senate

9:15 AM   **Session IV: University of Hawaii's Current Research
            Strengths and Security and Sustainability:
            Energy and Agriculture Opportunity**
            *Moderator: William Harris, President & CEO, Science Foundation
            Arizona*

            Hawaii's Satellite Launch Program
            *Brian Taylor, Dean School of Ocean and Earth Science and
            Technology, University of Hawaii at Mānoa*

Astronomy in Hawaii
Robert McLaren, Associate Director, Institute for Astronomy,
University of Hawaii

Data Analytics: A Proposal
The Honorable Daniel S. Goldin, Chairman & CEO, the Intellisis
Corporation and 9th NASA Administrator (Ret.)

Hawaii: A Model for Clean Energy Innovation
Maurice Kaya, Project Director, Hawaii Renewable Energy
Development Venture (HREDV)

Sustainable Agricultural Systems: Challenges and Opportunities
Sylvia Yuen, Dean, College of Tropical Agriculture and Human
Resources, University of Hawaii at Mānoa

11:15 AM **Lunch**

12:30 PM **Session V: Medical Opportunities in Hawaii**
Moderator: Virginia Hinshaw, Chancellor, University of Hawaii at
Mānoa

Clinical Trials in Hawaii
Art Ushijima, President and CEO, Queen's Health Systems/
The Queen's Medical Center

University of Hawaii Medical Initiatives
Jerris Hedges, Dean of John A. Burns School of Medicine,
University of Hawaii at Mānoa

Advancing Innovation and Convergence in Cancer Research
Jerry S. H. Lee, Deputy Director, Center for Strategic Scientific
Initiatives, National Cancer Institute, National Institutes of
Health

Biomedical Innovation with Global Impact in Hawaii
Hank Wuh, CEO, Skai Ventures and Cellular Bioengineering, Inc.

2:15 PM   **Session VI: Roundtable—Next Steps for Hawaii**
           Moderator: M.R.C. Greenwood, President, University of Hawaii

           Participants
           The Honorable Mazie Hirono, United States House of
               Representatives
           The Honorable Colleen Hanabusa, United States House of
               Representatives
           Peter Ho, Chairman, President and CEO, Bank of Hawaii, and
               Chairman, APEC 2011 Hawaii Host Committee
           Richard Rosenblum, President, Hawaiian Electric Company
           Donald Straney, Chancellor, University of Hawaii at Hilo
           Charles Wessner, Director, Technology, Innovation and
               Entrepreneurship, The National Academies

3:15 PM   **Adjourn**

Appendix B

Biographies of Speakers
*

NEIL ABERCROMBIE

The Honorable Neil Abercrombie grew up in Williamsville, New York, and attended Union College in New York State. He went to Hawaii in September 1959, the month after statehood, to be a teaching assistant at the University of Hawaii at Mānoa, where he earned a master's degree in sociology and later a Ph.D. in American studies. To support himself as a graduate student, Mr. Abercrombie worked as a waiter at Chuck's Steak House in Waikiki, a locker desk clerk at the Central YMCA, a custodian at Mother Rice Preschool, a construction apprentice program director, an elementary school teacher, and a college lecturer. Mr. Abercrombie served in the State House of Representatives from 1975 to 1979 and in the State Senate from 1979 to 1986. He helped elderly depositors regain their life savings when Mānoa Finance collapsed. Mr. Abercrombie helped start the award-winning Healthy Start prevention program for at-risk mothers and children. In 1986, Mr. Abercrombie won a special election to Congress to fill the remaining term of Congressman Cec Heftel. Mr. Abercrombie returned home to serve as special assistant to the Superintendent of Education where he helped implement Hawaii's distance-learning programs. Mr. Abercrombie was elected to serve on the Honolulu City Council from 1988 to 1990.

In 1990, Mr. Abercrombie was elected to return to Congress and represented Hawaii's 1st Congressional District in the U.S. House of Representatives until 2010. While in Congress, Mr. Abercrombie served as chairman of the Armed

---

* As of January 2011. Appendix includes bios distributed at the symposium.

Forces Subcommittee on Air and Land Forces, and as a senior member on the Natural Resources Committee.

Mr. Abercrombie is married to Dr. Nancie Caraway. They live in Lower Mānoa Valley with their dog, Kanoa, and their cat, Che.

DANIEL K. AKAKA

U.S. Senator Daniel Kahikina Akaka is America's first Senator of Native Hawaiian ancestry, and the only Chinese American member of the United States Senate. Like many of his generation, Senator Akaka's youth was interrupted by World War II. Upon graduation from high school, he served as a civilian worker then in active duty in the U.S. Army Corps of Engineers from 1943 to 1947. Following the war, Senator Akaka returned to school, enrolling in the University of Hawaii. A strong believer in the power of education, he made it his career, as a teacher and principal in the State of Hawaii Department of Education. First elected to the U.S. House in 1976, Congressman Akaka was appointed to the Senate when Senator Spark Matsunaga passed away, subsequently winning election to the office in 1990, and re-election in 1994, 2000, and 2006. Senator Akaka is chairman of the Veterans' Affairs Committee and the Homeland Security and Governmental Affairs Subcommittee on Oversight of Government Management, the Federal Workforce, and the District of Columbia. Senator Akaka also serves on the Armed Services, Indian Affairs and Banking, Housing and Urban Affairs Committees.

Raised in a deeply religious family, Senator Akaka is a member of the historic Kawaiahaʻo Church where he served as choir director for 17 years. He and his wife Millie are the parents of four sons and a daughter who have blessed them with 14 grandchildren and 10 great-grandchildren.

CARL BONHAM

Carl Bonham was a founding member of the University of Hawaii Economy Research Organization (UHERO) in 1999, and currently serves as UHERO's executive director and associate professor of economics. Dr. Bonham's research interests, include macroeconomics, applied dynamic econometrics, tourism economics, and the Hawaii economy. His current research includes development of dynamic factor models using daily, weekly, and monthly data on Hawaii's visitor industry to produce high-frequency forecasts of visitor arrivals and spending. Other work in progress includes a study of the effects of airfare on visitor travel decisions, and the impact of important agricultural land designations on Kauai. Recent publications include, "Modeling Tourism: A fully identified VECM approach," with Byron Gangnes and Ting Zhou in the *International Journal of Forecasting*. Dr. Bonham serves on the editorial board of the *Journal of Travel Research*, as a member of the State of Hawaii Council on Revenues, and on the

University of Hawaii President's Advisory Council on Hawaii Innovation and Technology Advancement.

Recent publications include: "Modeling Tourism: A fully identified VECM approach," with Byron Gangnes and Ting Zhou, *International Journal of Forecasting*, 25:531-49 (2009); and "Collusive Duopoly: The Effects of the Aloha and Hawaiian Airlines' Agreement to Reduce Capacity," with James Mak and Roger Blair, *Antitrust Law Journal*, 74(2):409-38 (2007).

KEIKI-PUA DANCIL

Keiki-Pua Dancil, Ph.D. is the president and Chief Executive Officer of Hawaii Science and Technology Council and Institute. Most recently Keiki-Pua was the executive vice president of Synedgen, Inc. (formally Hawaii Chitopure Inc), a diversified medical technology company. She also served as senior scientist and director of research and development for Trex Enterprises and its spin-off biosensor company, Silicon Kinetics. Keiki-Pua received her undergraduate degree from Santa Clara University in chemistry, her Ph.D. from the Department of Chemistry and Biochemistry from the University of California at San Diego, and her M.B.A. from Harvard Business School. She has several patents and has published in several journals, including *Science*.

MARY L. GOOD

Dr. Mary L. Good, founding dean and Donaghey Professor at the Donaghey College of Engineering and Information Technology of the University of Arkansas at Little Rock, is well known for her distinguished career. She has held many high-level positions in academia, industry, and government. The 143,000-member American Association for the Advancement of Science (AAAS) elected Dr. Good to serve as the president, following Dr. Stephen Jay Gould. In 2004, Dr. Good was recipient of the National Science Foundation's highest honor, the Vannevar Bush Award. She was also the first female winner of the AAAS's prestigious Philip Hogue Abelson prize for outstanding achievements in education, research and development management, and public service, spanning the academic, industrial, and government sectors. Two of her more than 27 awards include the National Science Foundation Distinguished Service medal and the esteemed American Chemical Society Priestly Medal. She is also the 6th Annual Heinz Award Winner. During the terms of Presidents Carter and Reagan, Dr. Good served on the National Science Board and chaired it from 1988 to 1991. She was the Undersecretary for Technology in the U.S. Department of Commerce and Technology during President Clinton's first term. This agency assists American industry to advance productivity, technology, and innovation in order to make U.S. companies more competitive in the global market.

Dr. Good has received 21 honorary degrees. Her undergraduate degree in chemistry is from the University of Central Arkansas. She earned her doctoral degree in inorganic chemistry from the University of Arkansas, Fayetteville, at age 24. Dr. Good spent 25 years teaching and researching at Louisiana State University and the University of New Orleans before becoming a guiding force in research and development for Allied Signal. Dr. Good was voted one of Arkansas' Top 100 Women by Arkansas Business.

M.R.C. GREENWOOD

An internationally known researcher and nationally recognized leader in higher education, M.R.C. Greenwood became the 14th president of the University of Hawaii (UH) in August 2009. She unites strong belief in the exceptional caliber of the UH system with determination to develop the university's voice as a national leader in higher education and research. The first woman to serve as UH's chief executive officer, Greenwood brings experience as both a campus and university system leader. She served as provost and senior vice president-academic affairs for the University of California (UC) system, focusing on long-range planning, graduate and professional education, a new science and math initiative, and more transparent admissions procedures. She previously served as chancellor of UC Santa Cruz during a period of great growth in academics programs, research initiatives, faculty hiring, and facilities development.

A national leader on science and technology policy and an expert on higher education policy issues, Greenwood served as associate director and consultant to the White House Office of Science and Technology Policy and chair of the National Academies Policy and Global Affairs Division. As a member of state and national committees and councils, she has dealt with issues from writing in America's schools and biomedical careers for women to national security and ethics of the information society. She is past-president and fellow of the American Association for the Advancement of Science, fellow of the American Academies of Arts and Sciences, member of the Institute of Medicine/National Academy of Sciences, and former member of the National Science Board. Greenwood has published extensively on education, obesity, diabetes, and women's health. The recipient of numerous scientific awards, she has been president of the Obesity Society and the American Society of Clinical Nutrition, chair of the Food and Nutrition Board of the Institute of Medicine, and fellow of the American Society for Nutrition.

She graduated summa cum laude from Vassar College and received her Ph.D. from The Rockefeller University. A self-described voracious and eclectic reader, she also enjoys volunteer work, hiking, sailing and being a "soccer grandma."

DANIEL S. GOLDIN

The Honorable Daniel S. Goldin is the founder, chairman, and CEO of The Intellisis Corporation, which develops neurobiologically inspired computational engines. Previously, as NASA's longest serving administrator from 1992 to 2001, he directly served three U.S. Presidents: George H.W. Bush, William Jefferson Clinton, and George W. Bush. Prior to NASA, he was the vice president and general manager of TRW Space and Technology Group where he oversaw a broad range of technology developments and programs for both government and industry. He began his career at NASA's Glenn Research Center working on electric propulsion systems for interplanetary travel. Mr. Goldin serves on the Board of Directors of AOptix Technologies and the Board of Trustees of the National Geographic Society. He is a member of the National Academy of Engineering, National Institutes of Health Scientific Management Review Board, Science and Technology in Society forum, International Academy of Astronautics, and Scripps Institute of Oceanography Advisory Council. He is a distinguished fellow at the Council on Competitiveness and a fellow of the American Institute for Aeronautics and Astronautics and the American Astronautical Society. He graduated from the City College of New York in 1962 with a B.S. in mechanical engineering.

COLLEEN HANABUSA

Congresswoman Colleen Hanabusa proudly represents Hawaii's 1st Congressional District. Congresswoman Hanabusa grew up in Waianae and graduated from the University of Hawaii at Mānoa where she received a bachelor of arts in economics and sociology, a master of arts in sociology, and a law degree from the William S. Richardson School of Law. She was elected into public office in 1998 to represent the people of the 21st District as their state senator. In the following years, she was elected vice-president of the State Senate and chaired several important committees. In 2006, she was elected Senate president, the first woman to lead either house of the Hawaii Legislature.

Along with her legislative work, Congresswoman Hanabusa is also an attorney with more than three decades of experience. She has been recognized by *Honolulu Magazine*'s as "One of Hawaii's A+ Attorneys," received the prestigious AV rating by Martindale-Hubbell, and was presented with the Spirit of Excellence Award by the American Bar Association Commission on Racial and Ethnic Diversity. Congresswoman Hanabusa is married to Honolulu businessman John Souza. They have a 7-year-old Border Collie named Little, who enjoys a farm fresh egg every morning for breakfast, specially prepared by John.

WILLIAM C. HARRIS

William Harris has more than 25 years of diverse, progressively advanced international experience building and leading major government and university institutions, strategically developing research and educational enterprises to benefit society, managing large budgets for maximum results, and hiring entrepreneurial personnel and teacher-scholars.

- President & CEO of Science Foundation Arizona (2006 to present). Responsible for initiation of a new public-private $235 million partnership to help the state of Arizona transform its focus on research and education.
- Director general, Science Foundation Ireland (2001-2006). Responsible for overall leadership of the new national science foundation and management of €635 million grants program. Built partnerships with numerous Irish government agencies to create dynamic academic-industry research centers, and recruited multinational corporate investment in these centers. Initiated the formation of effective national policies on technology transfer and intellectual property.
- Vice president for research and professor of chemistry, University of South Carolina, (2000-2001). Responsible for leading research activities totaling $125 million annually throughout the eight campuses and diverse interdisciplinary centers of the state's most comprehensive public university. Initiated a focus on interdisciplinary work by young faculty from the arts, humanities and social sciences. Established processes enabling students to earn support from incubator. Served on the Governor's Task Force, focused on building a more diverse, knowledge-based economy.
- Assistant director of the Columbia University Earth Institute (1996-2000) and president of the Biosphere 2 Center, the western campus of Columbia which focused on earth systems science and climate model testing. Developed an undergraduate semester system for students from about 30 colleges/universities. Led construction of a new campus complex.
- Assistant director for mathematical and physical sciences at the National Science Foundation (1992-1996). Responsible for the establishment of a major focus on interdisciplinary research by establishing the Office of Multidisciplinary Activities. Successfully introduced new materials focused group grants. Guided the successful establishment of a new magnet lab, the Gemini telescopes, and LIGO project.

JERRIS HEDGES

Dr. Jerris Hedges, dean of the John A. Burns School of Medicine since March 2008, is known around the country as co-author of one of the leading texts in patient care, *Roberts and Hedges' Clinical Procedures in Emergency Medicine,* now in its fifth edition.

Trained as an emergency physician, Dr. Hedges has spent the past 30 years contributing to the medical field through his work in clinical care, university teaching, research, and administration. At the internationally respected Oregon Health and Science University's (OHSU) School of Medicine, Dr. Hedges served as professor and department chair in emergency medicine and was named vice dean at OHSU in 2005. Dr. Hedges' leadership helped OHSU earn recognition as one of the "top 10" National Institutes of Health-funded academic emergency medicine research departments in the nation.

Dr. Hedges has also served as president of both the Society for Academic Emergency Medicine and the Association of Academic Chairs of Emergency Medicine. In October 2000, Dr. Hedges was elected to the prestigious National Academies' Institute of Medicine.

Dr. Hedges earned his bachelor's degree in aeronautics and astronautics, his master's degree in chemical engineering, and his medical degree at the University of Washington. He completed his residency at the Medical College of Pennsylvania and served on the faculty of the University of Cincinnati School of Medicine before joining OHSU. Dr. Hedges also holds a master of medical management from the Marshall School of Business at the University of Southern California. He and his wife, Susan, have two grown children.

VIRGINIA S. HINSHAW

Chancellor Virginia S. Hinshaw serves as the chief executive officer responsible for providing both administrative and academic leadership to the flagship campus of the University of Hawaii System.

Dr. Hinshaw is a renowned scientist with expertise in microbiology whose work over the past 25 years has contributed to the understanding of the influenza virus and new approaches to vaccines.

Prior to joining the University of Hawaii at Mānoa in 2007, Dr. Hinshaw served as the provost and executive vice chancellor at the University of California, Davis, and also served as the dean of the Graduate School and vice chancellor for research at the University of Wisconsin at Madison.

MAZIE K. HIRONO

Elected to Congress in 2006, Mazie K. Hirono is now serving her second term representing Hawaii's 2nd Congressional District in the U.S. House of Representatives. Hirono's district includes rural Oahu, the seven Neighbor Islands, and the Northwestern Hawaiian Islands. She serves on two key U.S. House committees: the Committee on Education and Labor and the Committee on Transportation and Infrastructure. She is an executive board member of the Congressional Asian Pacific American Caucus and a member of the House Democracy Assistance Commission.

Hirono's legislative priorities include education, transportation, renewable energy, and the environment. The Congresswoman has been recognized for her leadership in advancing quality early education for America's children. Pre-K Now, a national preschool advocacy organization, named Hirono its 2008 "Pre-K Champion" for her work and determination in passing the PRE-K Act (H.R. 3289) out of the Committee on Education and Labor on June 25, 2008.

Born in Fukushima, Japan, on November 3, 1947, Hirono became a naturalized U.S. citizen in 1959, the year Hawaii became a state. She is the first immigrant woman of Asian ancestry to be sworn into congressional office. Educated in Hawaii's public school system, Hirono graduated with honors from Kaimuki High School and Phi Beta Kappa from the University of Hawaii at Mānoa. She earned her law degree from Georgetown University Law Center in Washington, DC, where she focused on public interest law.

After graduation, she returned to Hawaii where she served as a deputy attorney general before entering private practice. In 1980, she was elected to the Hawaii State House of Representatives. In 1994, Hirono was elected and served two four-year terms as Hawaii's 9th lieutenant governor. Congresswoman Hirono is married to Leighton Kim Oshima.

PETER S. HO

Peter S. Ho is chairman, president, and chief executive officer of the Bank of Hawaii. Mr. Ho was named chairman and CEO on July 30, 2010. He has served as president since April 2008 and has been a member of the bank's board since December 2005. He was appointed to the parent company's board, Bank of Hawaii Corporation, in April 2009.

Mr. Ho joined the bank in 1993 as an assistant vice president in the National Banking Division. He was promoted to senior vice president in charge of corporate banking in 1999. In 2001, he was promoted to executive vice president responsible for corporate banking and commercial real estate lending. In 2003, Mr. Ho was promoted to group executive vice president in charge of the bank's Hawaii Commercial Banking Group and was made a member of the company's Managing Committee. In 2004, Mr. Ho was promoted to vice chairman responsible for the bank's Investment Services Group. In 2006, Mr. Ho was promoted to chief banking officer, responsible for both the commercial and investment services areas of the bank. In 2007, Mr. Ho was given the added responsibility for the organization's retail banking businesses. He was appointed president in April 2008.

Mr. Ho began his banking career in New York City in 1987. He holds a B.S. in business administration and an M.B.A. from the University of Southern California, where he was the 1992 First Interstate Bank Fellow. In 2008, Mr. Ho successfully attended and completed Harvard Business School's Advanced Management Program.

A native of Honolulu, Mr. Ho is the chairman of the APEC 2011 Hawaii Host Committee, the public-private entity created to support Hawaii's hosting of APEC Leaders Week in November 2011. He also serves on the boards of the Hawaii Chapter of the American Red Cross and University of Hawaii-Ahahui Koa Anuenue and on the board of trustees of McInerny Foundation and Strong Foundation. Mr. Ho was a 1998 Pacific Century Fellow and was named 2003 Pacific Business News Young Business Person of the Year. Mr. Ho is married (Michelle) and has two children.

DANIEL K. INOUYE

Daniel K. Inouye, the most senior member of the U.S. Senate and the president pro-tempore, is known for his distinguished record as a legislative leader, and as a World War II combat veteran with the 442nd Regimental Combat Team, who earned the nation's highest award for military valor, the Medal of Honor. Although he was thrust into the limelight in the 1970s as a member of the Watergate Committee and in 1987 as chairman of the Iran-Contra Committee, he has also made his mark as a respected legislator able to work in a bipartisan fashion to enact meaningful legislation.

As chairman of the Senate Appropriations Committee and of the Senate Defense Appropriations Subcommittee, Senator Inouye has been able to focus on defense matters that strengthen national security and enhance the quality of life for military personnel and their families. This reflects his hope for a more secure world and his desire to provide the best possible assistance to the men and women who put their lives at risk to protect the United States.

In addition, he is the ranking Democrat on the Commerce, Science & Transportation Committee and the Indian Affairs Committee and sits on the Rules Committee. He helped establish the Inter-parliamentary Exchange Program between the U.S. Senate and Japan's legislature, and in 2000 the Government of Japan presented him with the Grand Cordon of the Order of the Rising Sun. Early in his tenure in the Senate, Senator Inouye delivered the keynote address at the 1968 Democratic National Convention, and was under consideration to become Hubert Humphrey's vice-presidential running mate that same year. He became the first chairman of the Senate Select Committee on Intelligence in 1976 and served as the third-ranking leader among Senate Democrats as secretary of the Democratic Conference from January 1979 through 1988. He chaired the Senate Democratic Central America Study Group to assess U.S. policy and served as senior counselor to the National Bipartisan Commission on Central America (also known as the Kissinger Commission).

Senator Inouye has championed the interest of Hawaii's people throughout his career. With his support, Hawaii's infrastructure has been strengthened, its economy diversified, and its natural resources protected and restored. For local residents, particularly Native Hawaiians, whose history and welcoming culture give the state its defining characteristics, Senator Inouye has increased job training and employment opportunities, provided more community health care, and provided support services and research to help small businesses and diverse sectors, from agriculture to high technology.

His imprint is seen on all of the state's islands through initiatives such as Honolulu and Neighbor Island bus service, steady construction jobs in support of military infrastructure, the diversification of agriculture, the birth of the Kauai High Technology Center and the rise of the Pacific Missile Range Facility, the launch of the Maui supercomputer, the expansion of national parks and wildlife refuges in Hawaii, and the protection of Hawaiian monk seals, sea turtles, the alala (Hawaiian crow), the nene goose, and coral reefs.

Senator Inouye got his start in politics in 1954 when he was elected to the Territorial House of Representatives; soon after his election, his Democratic colleagues, well aware of Inouye's leadership abilities, selected him as their majority leader. In 1958 he was elected to the Territorial Senate. When Hawaii became a state in 1959, he was elected the first Congressman from the new state, and was re-elected to a full term in 1960. He was first elected to the U.S. Senate in 1962 and is now serving his eighth consecutive term.

On May 24, 2008, Senator Inouye married Irene Hirano, who is President of the U.S.-Japan Council. He was married for nearly 57 years to Margaret Awamura Inouye, a former instructor at the University of Hawaii, who passed away on March 13, 2006. He has a son, Ken, who is married to Jessica Carroll from Rochester, New York, and a granddaughter Mary Margaret "Maggie" Inouye.

MAURICE KAYA

Maurice Kaya joined Hawaii Renewable Energy Development Venture in 2008; as project director he is responsible for the strategic direction and overall execution of the project. Mr. Kaya has over 35 years of experience in energy and environmental engineering in Hawaii's public and private sectors.

Mr. Kaya served as the director of the State of Hawaii's energy program (1988-2008) and was also appointed chief technology officer (CTO) for the Hawaii Department of Business, Economic Development and Tourism (DBEDT) (2003-2008). As energy program director and CTO, he was responsible for the planning and execution of state energy policy and projects.

Mr. Kaya also developed a comprehensive energy policy strategy for the State of Hawaii, which was implemented first through Energy for Tomorrow and is now part of the Hawaii Clean Energy Initiative, a partnership between the state and the U.S. Department of Energy that he helped conceive and launch. He has served on numerous boards dealing with energy and high-technology development. He currently serves as a board member of Energy Industries and the U.S. Department of Energy (DoE) Hydrogen and Fuel Cell Technical Advisory Committee, as well as co-chair of the Hawaii EPSCoR Statewide Committee.

Previously, he served as the chairman of the State Energy Advisory Board to the Secretary of Energy, U.S. DoE, and on the boards of the National Association of State Energy Officials, the Natural Energy Laboratory of Hawaii Authority, and the High Technology Development Corporation. He also served as the chairman of the Emerging Energy Technologies technical committee, EnergyDivision, American Society of Civil Engineers. Mr. Kaya's prior positions include vice president and chief engineer of a major civil engineering firm in Honolulu, the deputy director and chief engineer for the City and County of Honolulu, and the director of Facilities Planning and Engineering, U.S. Navy Public Works Center, Pearl Harbor.

Mr. Kaya received his B.S. in civil engineering and M.S. in environmental engineering from the University of Hawaii at Mānoa.

ROGER D. KILMER

Roger Kilmer is the director of the Hollings Manufacturing Extension Partnership (MEP), a program of the Department of Commerce's National Institute of Standards and Technology (NIST).

Mr. Kilmer has been with the MEP program since 1993 and with NIST since 1974. Previously, Mr. Kilmer was the MEP deputy director, serving as the chief operating officer and chief financial officer responsible for internal operations, programmatic coordination, and policy review of all activities. From 1990 to 1993, Mr. Kilmer was the deputy division chief of Robot Systems in the NIST Manufacturing Engineering Laboratory. In this position, he was responsible for establishing and managing research programs involving real-time sensor-based control of intelligent machines. Mr. Kilmer was also group leader of Robot Systems Integration, managing research and development programs with manufacturing and military applications including robotic deburring, automated lay up of thermoplastic composites, robotic safety systems, robotic handling of munitions, and unmanned land vehicle operations.

Mr. Kilmer received the Department of Commerce Gold Medal Award for the CommerceConnect initiative, the Silver Medal Award for leadership as the NIST MEP liaison to the interagency Technology Reinvestment Project (TRP) initiative, and the Bronze Medal Award for superior leadership of NIST's unmanned ground vehicle robotics program.

Mr. Kilmer holds a M.S. and a B.S. in mechanical engineering from Pennsylvania State University.

BARRY E. A. JOHNSON

Barry Johnson serves as senior advisor and director of strategic initiatives for the U.S. Economic Development Administration (EDA) within the U.S. Department of Commerce.

With over 20 years of business experience as an entrepreneur and corporate executive with a focus on public-private partnerships, Barry leads EDA's strategic partnership programs, collaborating with the White House, other federal agencies, and regional leaders to implement innovative regional solutions to improve competiveness and foster economic growth throughout the nation.

Prior to this role, he was founder and managing principal at Acresh Development, a public-private real estate development and advisory firm. Barry also has an extensive background in starting innovative new businesses within media giants such as the Walt Disney Company and Sony Music Corporation. He was formerly founding president of MSBET, a joint venture between Microsoft Corporation and BET Holdings. Committed to giving back through volunteerism, Barry is an active mentor and motivational speaker on college campuses nationally.

A native of Birmingham, Alabama, Barry earned an M.B.A. from the Harvard Business School after receiving B.S. degrees in economics and political science from Yale College.

KATHERINE KU

Katharine Ku is director of the Office of Technology Licensing (OTL) at Stanford University. OTL is responsible for the licensing of various state-of-the-art university technologies and industry sponsored research agreements and collaborations. In fiscal year 2003-2004, OTL received $49.5 million from the licensing of over 435 different technologies. From 1994 to 1998, in addition to her OTL responsibilities, Ku was responsible for Stanford's Sponsored Projects Office, which handled $500 million in research contracts and grants. Ku was vice president, business development, at Protein Design Labs, Inc. in Mountain View, California from 1990 to 1991. Prior to PDL, Ku spent 12 years at Stanford in various positions, was a researcher at Monsanto and Sigma Chemical, administered a dialysis clinical trial at University of California, and taught chemistry and basic engineering courses.

Ku has been active in the Licensing Executive Society (LES), serving as vice president (Western Region), trustee, and various committee chairs. She also has served as president of the Association of University Technology Managers (AUTM) from 1988 to 1990. She received the AUTM 2001 Bayh-Dole Award for her efforts in university licensing.

Ku has a B.S. in chemical engineering (Cornell University) and an M.S. in chemical engineering (Washington University) and is a registered patent agent.

JERRY S. H. LEE

Dr. Lee serves as the deputy director for the National Cancer Institute's (NCI's) Center for Strategic Scientific Initiatives (CSSI). He provides scientific input and expertise to the planning, coordination, development, and deployment of the innovation center's strategic scientific initiatives. Dr. Lee serves and leads various trans-NCI working groups and also represents CSSI at various NIH, Health and Human Services (HHS), and external committees and other activities to develop effective partnerships across federal agencies, and to build collaborations with key external stakeholders.

Dr. Lee is responsible for providing day-to-day administrative and programmatic management for CSSI's offices including: (1) The Cancer Genome Program Office (TCGA PO); (2) Office of Cancer Nanotechnology Research; (3) Office of Biorespositories and Biospecimen Research (OBBR); (4) Office of Cancer Genomics (OCG); (5) Office of Cancer Clinical Proteomics Research (OCCPR); and (6) Office of Physical SciencesOncology (OPSO). He serves as acting director for the Office of Physical Sciences-Oncology, responsible for initiatives at the interface of physical and life sciences including the NCI's Physical Sciences-Oncology Centers (PS-OCs) program, and also as acting director for the TCGA Program Office.

Dr. Lee's efforts facilitate the execution of cross-disciplinary strategies and synergies in key areas of research and training to support these emerging fields. His past experience at NIH includes serving as a program manager for the NCI's Innovative Molecular Analysis Technologies (IMAT) program and the NCI Alliance for Nanotechnology in Cancer program, where he was program director of fellowships to support multidisciplinary training in cancer nanotechnology. Dr. Lee's previous research experiences in coordinating collaborations among the Naval Research Laboratory, NCI-Frederick Laboratory, JHU Medical Oncology Division, and the Institute for NanoBioTechnology also contribute to carrying out his current efforts.

Scientifically, Dr. Lee has extensive research experience in using engineering-based approaches to examine mechanisms of age-related diseases and cancer progression focused on combining cell biology, molecular biology, and engineering to understand various cellular reactions to external stimuli. Specifically, Dr. Lee's research has emphasized increasing the understanding of RhoGTPase-mediated nuclear and cellular mechanical responses to fluid flow, 3D culture, and contributions to laminopathies such as progeria. He has coauthored numerous papers, two book chapters, and one book, and has spoken at various cell biological and biomedical conferences.

Dr. Lee currently serves as adjunct assistant professor at Johns Hopkins University, where he also earned his bachelor's degree in biomedical engineering and Ph.D. degree in chemical and biomolecular engineering.

GINGER LEW

Ginger Lew is senior counselor to the White House National Economic Council and the Small Business Administration (SBA) Administrator. She provides economic policy advice on a broad range of matters that impact small businesses.

In addition, she co-chairs the White House Interagency Group on Innovation and Entrepreneurship. Prior to joining the Obama Administration, Ms. Lew was the managing partner of a communications venture capital fund, and a venture advisor to a Web 2.0 venture fund.

Under the Clinton Administration, Ms. Lew was the deputy administrator and chief operating officer of the Small Business Administration where she provided day-to-day management and operational oversight of a $42 billion loan portfolio. Before joining SBA, Ms. Lew was the general counsel at the U.S. Department of Commerce where she specialized in international trade issues. Ms. Lew was unanimously confirmed by the United States Senate for both positions.

For the past 10 years, Ms. Lew was chairman and board member of an investment fund based in Europe. She has served on the boards of publicly traded companies, private companies, and nonprofit organizations.

ROBERT MCLAREN

Bob McLaren grew up in the small town of Watford, in southwestern Ontario (Canada). He studied physics at the University of Toronto, obtaining a Ph.D. in the field of laser spectroscopy in 1973. He then spent two years at the University of California at Berkeley as a NATO postdoctoral fellow. During this period, he reoriented his research interests from laboratory physics to infrared astronomy. In 1975 he returned to Toronto to take up a faculty position in the Department of Astronomy. From 1982 to 1990, he held a series of positions at the Canada-France-Hawaii Telescope, culminating in service as its executive director.

In 1990, Dr. McLaren joined the faculty of the University of Hawaii Institute for Astronomy. His main work has been the implementation of the university's plan for the astronomical development and utilization of Mauna Kea. This involves the characterization and preservation of the superb qualities of the Mauna Kea site, liaison with existing and proposed new telescope facilities, and the planning and execution of infrastructure improvements. Dr. McLaren served as interim director of the institute from July 1997 through September 2000. Since then, he has held the position of associate director. He continues his work related to the Mauna Kea Observatories and teaches introductory astronomy.

Dr. McLaren is a member of the American Astronomical Society and the International Astronomical Union. He currently serves on the boards of the Canada-France-Hawaii Telescope Corporation and the Gemini Observatory.

DANIEL T. OLIVER

Daniel T. Oliver was commissioned in the Navy in 1966 and spent his operational career as a P-3 aviator, rising to command a squadron and a patrol wing. He has served on the staffs of two Chiefs of Naval Operations, and was a White House fellow. As a Flag Officer, President Oliver has served as Commander,

Fleet Air Forces Mediterranean, various tours in the Pentagon in resourcing, planning and budgeting, and finished his very successful 34-year career as the Chief of Naval Personnel and Deputy Chief of Naval Operations for Manpower and Personnel.

After retiring from active duty in February 2000, he was active in the private sector as a senior executive and board member of a number of companies and civic organizations. In April 2007, he accepted an offer from the Secretary of the Navy to lead the Naval Postgraduate School.

LUIS M. PROENZA

Luis M. Proenza is president of The University of Akron and an experienced leader in national science and technology policy matters. Prior to his appointment at Akron, he was then vice president for research and dean of the Graduate School at Purdue University and previously vice chancellor for research and dean of the Graduate School and then vice president for academic affairs and research at the University of Alaska. Dr. Proenza served on the U.S. Arctic Research Commission (U.S. Presidential appointment); Advisory Board of the U.S. Secretary of Energy, chairing the Science and Mathematics Education Task Force; NAS-NRC Committee on Vision; National Biotechnology Policy Board; and as advisor for science and technology policy to Alaska's governor. In 2001, the President of the United States appointed Proenza to the President's Council of Advisors on Science and Technology (PCAST), the nation's highest-level policy-advisory group for science and technology. His PCAST panel work included U.S. research and development investments, technology transfer, energy efficiency and advanced manufacturing, nanotechnology, alternative energy, and information technology. Proenza is on the executive committee and the National Innovation Initiative Leadership Council of the Council on Competitiveness, co-chairs its Regional Leadership Institute Steering Committee, and serves on the Steering Committee for the Energy Security, Innovation and Sustainability Initiative.

He also is on the Council on Foreign Relations, The National Academies' Government-University-Industry Research Roundtable, the Technology Innovation Program Advisory Board for the National Institute of Standards and Technology, and the board of the States Science and Technology Institute, and he is Association of Public and Land-Grant Universities co-chair of the APLU/AAU Patent Reform Committee.

After earning a B.A. from Emory University (1965), M.A. from The Ohio State University (1966), and Ph.D. from the University of Minnesota (1971), Dr. Proenza joined the faculty of the University of Georgia. There his research was continuously supported by grants from the National Eye Institute, including a Research Career Development Award, and he served as assistant to the president and university liaison for science and technology policy.

RICHARD M. ROSENBLUM

Dick Rosenblum was named president and chief executive officer of Hawaiian Electric Company, Inc. (HECO) effective January 1, 2009, and on February 23, 2009, he was appointed a director of the HECO Board. He had retired on May 1, 2008, as senior vice president of generation and chief nuclear officer for Southern California Edison (SCE), responsible for all power generating facilities, including nuclear and related fuel supplies. He was appointed to this position in November 2005.

Previously, Mr. Rosenblum was senior vice president of SCE's transmission and distribution business unit which is responsible for the high-voltage bulk transmission and retail distribution of electricity in SCE's 50,000 square mile service territory. He held that position since February 1998. Prior to that, he was vice president of the distribution business unit, responsible for providing electric service to SCE's 4.6 million customers.

Mr. Rosenblum began his career at SCE in 1976 as an engineer working at the company's San Onofre Nuclear Generating Station. He held various positions in the company's Nuclear Department and was named vice president of engineering and technical services in 1993. In that role he was responsible for engineering construction, safety oversight, and other engineering support activities. Mr. Rosenblum is on the boards of the High Technology Development Corporation, the Hawaii Employers Council, the Aloha Council, and Boy Scouts of America, and he is a member of the Hawaii Business Roundtable. In addition, he was the 2010 corporate recruitment chair for the American Diabetes Association's annual Step Out to Fight Diabetes Walk. Mr. Rosenblum earned a B.S. and M.S. in nuclear engineering from Rensselaer Polytechnic University. He and his wife, Michele, have two grown children and two grandchildren.

BRIAN SCHATZ

Lieutenant Governor Brian Schatz was raised in Hawaii and has devoted his life to public service. He is known for his energy, compassion, and problem-solving skills. Brian was a member of the State House of Representatives for four terms. During this time, he served as the House majority whip and chair of the Economic Development Committee. Brian served for eight years as the chief executive officer of a major human services agency, Helping Hands Hawaii.

Working with a dedicated staff and using a determined and collaborative leadership style, Brian successfully led the agency through a difficult financial period. Today it serves many of Hawaii's most needy people. Brian put his values into action by starting the Hawaii Draft Obama campaign in 2006, helping to elect the first President of the United States born in Hawaii. He also served as the chair of the Democratic Party of Hawaii. During this period, membership in the Democratic Party more than doubled.

Brian is married to Linda Kwok Schatz and they have two children. He is a devoted husband, father, and public servant. Brian is committed to working with Governor Abercrombie to bring positive change to the State of Hawaii.

DONALD O. STRANEY

Donald O. Straney took up the position of University of Hawaii at Hilo chancellor on July 1, 2010. Previously, Dr. Straney served as dean of science at California State Polytechnic University, Pomona, where he was also professor of biological sciences. He had joined Cal Poly Pomona in 2002 after spending 23 years at Michigan State University, where he served as chair of the Department of Zoology from 1986 to 1995 and as assistant to the provost for faculty development from 1995 to 2002.

Dr. Straney is on the National Advisory Board of the National Science Foundation (NSF)-supported Center for the Integration of Teaching, Research and Learning at the University of Wisconsin. He has been a principal investigator for three large grants at Cal Poly Pomona: a Howard Hughes Medical Institute grant to enhance undergraduate instruction in biology, an NSF ADVANCE grant to support the professional development of science and engineering faculty, and a U.S. Department of Education Teacher Quality Enhancement grant to prepare the next generation of teachers. He also led the university's efforts to establish a twinning program in biotechnology, computer science, business, and mechanical engineering with Technology Park Malaysia College, a new institution in Kuala Lumpur, and with Al Akhwayn University in Morocco.

Within the California State University system, he served on the board of directors of both the Desert Studies Center and the Ocean Studies Institute as well as on the Strategic Planning Council of CSUPERB, the California State University Program for Education and Research in Biotechnology. An evolutionary biologist by training, Dr. Straney has studied patterns of change in a variety of organisms, most recently focusing on ants. He received a Ph.D. in zoology from the University of California, Berkeley, and both his M.S. and B.S. degrees are from Michigan State University in zoology.

BRIAN TAYLOR

Dr. Brian Taylor is dean of the School of Ocean and Earth Science and Technology (SOEST) at the University of Hawaii at Mānoa. SOEST is a $140 million/year operation with about 850 employees, including 230 Ph.D.s, 440 staff, and 180 graduate assistants. The school is an international leader in such diverse fields as alternative energy, tropical meteorology, coral reef ecosystems, volcanology, microbial oceanography, seafloor processes, hyperspectral remote sensing, cosmochemistry, coastal processes, and climate modeling—and that is just the top 10.

With a B.Sc. Hons.(1st) from the University of Sydney and a Ph.D. from

Columbia University in marine geology and geophysics, Brian's background is in plate tectonics and seafloor volcanism, deformation, and sedimentation. A former Fulbright Fellow and JOI Distinguished Lecturer, he is the treasurer of the Board of Trustees of the Consortium for Ocean Leadership, and chairman of the Board of Governors of the Integrated Ocean Drilling Program—Management International.

TYRONE C. TAYLOR

Tyrone C. Taylor is the founder and president of Capitol Advisors on Technology, LLC, located in Washington, DC. He is a former member of the Senior Executive Service at NASA where he worked on a variety of programs including Space Science and Space Station, and served as the agency's lead on technology transfer issues. As an executive on loan he served as the Washington, DC, Representative for the Federal Laboratory Consortium, a congressionally chartered organization, representing the nation's defense and non-defense laboratories in the area of technology transfer. While working in the private sector, Mr. Taylor has provided technology and management support to a variety of federal agencies including the National Science Foundation, NASA, and the Department of Defense and private-sector firms such as General Electric on a broad spectrum of technology issues including homeland security, innovation management and commercialization of space as examples. He is the former chair, Small Business Committee, National Defense Industrial Association, and has served on numerous advisory committees.

ART USHIJIMA

Art Ushijima is the president and CEO of The Queen's Health Systems and president of The Queen's Medical Center, Hawaii's largest adult tertiary care teaching hospital affiliated with the University of Hawaii's John A. Burns School of Medicine. He been at Queen's for the past 22 years; prior to Queen's, he has served in senior management roles in four other community and teaching hospitals in Arizona, Nebraska, Missouri and Ohio. He earned his M.A. in hospital and health care administration from the University of Iowa.

STARNES WALKER

Dr. Starnes Walker is the chief engineering and technical director at the University of Hawaii. Until his hire by the University of Hawaii in December 2010, Dr. Walker served as the director of research in the Science and Technology (S&T) Directorate of the U.S. Department of Homeland Security (DHS) in Washington, DC, where he oversaw the Office of National Laboratories, the Office of University Programs for the DHS Centers of Excellence, and the Academic Fellowship and Scholarship Program Office. Dr. Walker joined the S&T

Directorate in January 2007 from the Office of Naval Research (ONR), where he was technical director/chief scientist for the Naval S&T program. He also served as the technical director and chief scientist reporting directly to the Chief of Naval Research (CNR). Working with the CNR, Dr. Walker was responsible for structuring and leading an S&T organization that ensures technological superiority for the U.S. Navy and Marine Corps.

Dr. Walker's budget authority was annually $2.2 billion, plus an additional average congressional increase of $700 million, and Dr. Walker supervised a workforce of 5,500 civilian and military personnel for ONR and the Naval Research Laboratory. Dr. Walker's leadership spanned the university community, the government laboratory structure, industry, and international government defense organizations to bring their resources and technical capabilities into the Naval S&T program, thereby ensuring strategic Naval capabilities to the future and avoiding technological surprise for the nation. Dr. Walker's previous position was as the acting associate laboratory director for national security, serving as the national security coordinator at Argonne National Laboratory. Most recently, Dr. Walker served on the DoD's Defense Science Board in the Summer Study to define Future Strategic Strike Systems with STRATCOM as the COCOM sponsor. Dr. Walker is a former member of the Senior Executive Service and served as the senior advisor for science and technology at the Defense Threat Reduction Agency from 2000 to 2003.

Dr. Walker was a standing member of the Defense Science and Technology Advisory Group for DDR&E in the Office of the Secretary of Defense. He started his career at the Naval Weapons Center-Corona Laboratories in 1968 as a research physicist. In 1970 he joined the Naval Weapons Center-China Lake. In 1973, Dr. Walker joined Phillips Petroleum as a research physicist. Advancing to a senior scientist position, he founded and directed programs in physics, technology, nuclear weapons support, energy, and bioengineering, as well as an ending assignment serving as the environmental director for operations. From 1992 to 1998, he served as vice president for technology for Morrison Knudsen Corporation with responsibility for developing new technology and engineering partnerships with the DoD, state, and national laboratories. From 1998 to 1999, Dr. Walker led a team with British Nuclear Fuels Limited that successfully developed a new process from an R&D platform through pilot plant demonstration for the chemical separation of transuranics. Dr. Walker has B.S., M.S., and Ph.D. degrees in physics from the University of California.

He has an honorary degree in nuclear engineering from the University of Missouri-Rolla. Dr. Walker is chairman of the Joint Laboratory Board of the Joint Improvised Explosive Devices Defeat Organization (JIEDDO), previously known as the JIEDD Task Force. He also serves as chairman of the Engineering Development Board of the University of Missouri-Rolla and as a guest scientist to Los Alamos National Laboratory.

Previously, Dr. Walker served as science advisor to Lawrence Livermore

National Laboratory and led a tritium production R&D program at the Idaho National Engineering Laboratory. As a distinguished member and senior advisor for S&T, Dr. Walker was awarded in 2002 the DTRA Exceptional Civilian Service Medal. He received the R&D 100 Award in 1980, and he has served on the Air Force Studies Board, National Academy of Sciences Committees, and Institute of Chemical Waste Management Steering Committee. His team, for their leadership in Project Sapphire, received a Presidential Citation from the White House. Dr. Walker has widely published in the fields of physics, chemistry, and optics, with numerous patents issued. He was a Navy fellow and recipient of three consecutive Naval Weapons Fellowship awards. Dr. Walker is a member of the American Physical Society and American Nuclear Society.

MARY WALSHOK

Mary Lindenstein Walshok, Ph.D., a sociologist, is associate vice chancellor and dean of the Extension Division at the University of California, San Diego. Over three decades, she has been a catalyst in building regional collaborations focused on high-tech cluster development (UCSD CONNECT) and cross-border synergies (the San Diego Dialogue) based on San Diego's proximity to Mexico. She is the author of four books: *Blue Collar Women*, *Knowledge Without Boundaries*, *Closing America's Job Gap*, and *Invention and Reinvention: The Evolution of San Diego's Innovation Economy*, forthcoming in Stanford University Press. She has also authored more than 100 reports and articles on the regional competencies and social dynamics essential to building knowledge-based clusters and high-wage jobs. Walshok's current research activities include serving as the principal investigator for the evaluation of 13 Generation I WIRED regions funded by the U.S. Department of Labor; a two-year NSF-funded project comparing the distinctive social dynamics of three innovation regions—Philadelphia, St. Louis, and San Diego; an NIH-funded comparative study of research outcomes in Central Florida, the DC/Baltimore corridor, and San Diego; and a Lilly Foundation-funded assessment of efforts to sustain and grow the robust orthopedic device industry in Warsaw, Indiana.

Walshok is the recipient of numerous awards including the distinguished Kellogg Foundation's Leadership Fellowship and, most recently, induction into Sweden's Royal Order of the Polar Star. Active on boards of a number of arts and philanthropic organizations, Walshok chaired the boards of the San Diego Community Foundation from 2002 to 2004 and the International Community Foundation from 2007 to 2009. She is currently serving on the boards of the San Diego CONNECT, the La Jolla Playhouse, the United States-Mexico Foundation for Science, International Community Foundation, and the Girard Foundation.

BARRY WEINMAN

Barry Weinman has been a venture capitalist since 1980. He is founder of Allegis Capital, a Palo Alto-based venture fund with over $700 million under management and a Red Herring Venture 100 Firm (#28) out of 1,800 global venture firms. Mr. Weinman has led and participated in investments resulting in over $35 billion in market value, including: Palm (NASDAQ), Cypress Semiconductor (NYSE), and Columbia/HCA (NYSE). Mr. Weinman has a B.S. from Clarkson College of Technology and an M.A. from the University of Southern California (USC)/London School of Economics.

As a U.S. Navy officer he was a speech writer and briefing officer for Admiral John McCain, the Commander of the U.S. Naval Forces Europe, and David Bruce, U.S. Ambassador to London. From 1989 to 1995, he was a lecturer on entrepreneurship at the USC Business School. He is chairman of the board of trustees of the University of Hawaii Foundation (Endowment) and was the chair of the University of Hawaii Centennial Campaign, which raised $336 million against a goal of $250 million.

CHARLES W. WESSNER

Charles Wessner is a National Academy Scholar and director of the Program on Technology, Innovation, and Entrepreneurship. He is recognized nationally and internationally for his expertise on innovation policy, including public-private partnerships, entrepreneurship, early-stage financing for new firms, and the special needs and benefits of high-technology industry. He testifies to the U.S. Congress and major national commissions, advises agencies of the U.S. government and international organizations, and lectures at major universities in the United States and abroad. Reflecting the strong global interest in innovation, he is frequently asked to address issues of shared policy interest with foreign governments, universities, research institutes, and international organizations, often briefing government ministers and senior officials. He has a strong commitment to international cooperation, reflected in his work with a wide variety of countries around the world.

Dr. Wessner's work addresses the linkages between science-based economic growth, entrepreneurship, new technology development, university-industry clusters, regional development, small-firm finance, and public-private partnerships. His program at the National Academies also addresses policy issues associated with international technology cooperation, investment, and trade in high-technology industries.

Currently, he directs a series of studies centered on government measures to encourage entrepreneurship and support the development of new technologies and the cooperation between industry, universities, laboratories, and government to capitalize on a nation's investment in research. Foremost among these is a congressionally mandated study of the Small Business Innovation Research (SBIR) Program, reviewing the operation and achievements of this $2.3 billion award program for small companies and start-ups. He is also directing a major study on best practice in global innovation programs, titled *Comparative National Innovation Policies: Best Practice for the 21st Century*. Today's meeting "E Kamakani Noi'i" forms part of a complementary analysis entitled *Competing in the 21st Century: Best Practice in State and Regional Innovation Initiatives*. The overarching goal of Dr. Wessner's work is to develop a better understanding of how we can bring new technologies forward to address global challenges in health, climate, energy, water, infrastructure, and security.

HANK C. K. WUH

An orthopedic surgeon, inventor, and entrepreneur, Dr. Hank C. K. Wuh has led the development of over twenty biomedical and consumer health care products from concept to global commercialization.

Dr. Wuh is founder and CEO of Skai Ventures, a hybrid of venture capital and technology accelerator focused on developing successful companies by transforming novel, ingenious ideas from scientists at leading universities into disruptive innovations. He is also founder and CEO of Cellular Bioengineering, Inc., developing the world's most advanced, bioengineered cornea for transplantation to restore vision for the 10 million people around the world with corneal blindness (*www.cellularbioengineering.com*). Dr. Wuh recently founded World Children's Vision, a charity with a mission to bring blind children from around the world to Hawaii for the gift of sight.

Skai Ventures' portfolio companies include DeconGel® (*www.decongel.com*), a polymeric material for radiological, nuclear, and hazardous chemical remediation; TruTags (*www.TruTags.com*), an edible optical security platform targeting the $75 billion annual problem of pharmaceutical counterfeiting; International Center of Excellence for Vision, delivering leading-edge technology to restore sight for visually impaired patients from around the world; and StemPure, optimizing the safety of stem cell transplantation.

Dr. Wuh received his B.A. from Johns Hopkins, M.P.H. from Harvard, and M.D. from the Johns Hopkins School of Medicine. Dr. Wuh was resident and chief resident in orthopedic surgery at the Stanford Medical Center. He was elected class marshal at Harvard and was twice winner of the Vernon P. Thompson Prize for outstanding research in orthopedic surgery at Stanford. Dr. Wuh was named the 2008 Invention Entrepreneur of the Year by the Hawaii Venture Capital Association and was nominated in 2010 for The National Medal of Technology and Innovation.

At the University of Hawaii, Dr. Wuh is a member of the President's Advisory Council on Innovation and Technology Advancement and the Dean's Council at the School of Engineering, and he is associate clinical professor of surgery at the John A. Burns School of Medicine. Dr. Wuh is board director of the Hawaii Science and Technology Council (HSTC) and the Hawaii Business and Entrepreneur Acceleration Mentors (HiBEAM).

SYLVIA YUEN

Dr. Yuen is the interim dean and director of the College of Tropical Agriculture and Human Resources (CTAHR) at the University of Hawaii (UH), the first woman to lead the college. She previously served as the director of the Center on the Family, CTAHR's associate dean for academic affairs, UH Mānoa director of Equal Employment Opportunity, and UH's employee relations administrator. Her graduate work was conducted at the University of Illinois and the University of Hawaii, and she received additional training at the University of Chicago, the University of Michigan, and Harvard University.

Dr. Yuen serves as the PI of the Agricultural Development in the American Pacific (ADAP) program, a consortium of land-grant institutions in the Western Pacific. Her publications have targeted both professional and community audiences, and she and her colleagues have won awards for a data-based Web site, a videotape on families, and other work. She is the recipient of both CTAHR's and Maryknoll High School's Outstanding Alumnus Awards, three Excellence in Teaching awards, more than $22 million in grants and contracts, and commendations from the Hawaii State Legislature for distinguished service to the state.

Appendix C

Participants List[*]

Piia Aarma
Pineapple Tweed PR & Marketing

Neil Abercrombie
Governor of the State of Hawaii

Lori Admiral
University of Hawaii Foundation

Palmer Ahakueo
Honolulu Community College

Daniel K. Akaka
United States Senate

Ryan Akamine
City Council

Bill Akiona, II
OmniGreen Renewables

Terri Alvaro
University of Hawaii Foundation

Keith Amemiya
University of Hawaii Board of Regents

Dana Anderson
OLA Hawaii 2020

Henry Aquino
Hawaii State House of Representatives

Russell Au
HTDC

Krystyna Aune
UHM-OVCAA

Amita Aung-Thwin
State of Hawaii

Kent Avary

Gene Awakuni
University of Hawaii at West Oahu

Karen Awana
Hawaii State House of Representatives

Kristi Bates
University of Hawaii Foundation

Artemio Baxa
University of Hawaii Board of Regents

Brian Bell

Edoardo Biagioni
University of Hawaii at Mānoa

Kim Binsted
University of Hawaii

Andre Bisquera
University of Hawaii

Kimo Blaisdell
Queen's Health Systems

Carl Bonham
University of Hawaii Economy
 Research Organization (UHERO)
University of Hawaii at Mānoa

Kristen Bonilla
University of Hawaii

Bernice Bowers
Inovi Group LLC

Robin Brandt

Stephen Brennan
Concentris Systems LLC

Mangmang Brown
University of Hawaii Foundation

Peter Bryant-Greenwood
Queen's Medical Center

Janet Bullard
University of Hawaii Foundation

Jennifer Burke

Manny Cabral
Leeward Community College

Tom Cannon
Tissue Genesis, Inc.

Howard Carr
Board of Regents
University of Hawaii

Stanford Carr

John Carroll
Omega GS

Paul Casey

Kara Catlin

Sun-Ki Chai
University of Hawaii

Paul Chandler
University of Hawaii at Mānoa

Jerry Chang
Hawaii State House of Representatives

Cheryl Chappell-Long
University of Hawaii Community
 Colleges

Kymber Char
University of Hawaii

---

* Speakers in italics.

Russel Cheng

Stanley Ching

Bill Chismar
University of Hawaii Outreach College

John Chock
University of Hawaii Shidler College
    of Business

Pia Chock
Hawaii Small Business Development
    Center

Johnson Choi
Hong Kong/China/Hawaii Chamber
    of Commerce

Song Choi

Malia Chow
NOAA Hawaiian Islands Humpback
    Whale National Marine Sanctuary

Devin Choy

Bee Leng Chua
HiBEAM

Harriet Cintron
University of Hawaii Foundation

McAlister Clabaugh
The National Academies

Stuart Coleman
Surfrider Foundation
University of Hawaii English
    Department

Patricia Cooper
University of Hawaii at Mānoa

Helen Cox
Kaua'i Community College

Stephen Craven
Kekepana International Services

Rolanse Crisafulli
Oahu WorkLinks

Peter Crouch
University of Hawaii

Douglas Cullison

Mike Curtis

Kathy Cutshaw
University of Hawaii

Michael Dahilig
University of Hawaii Board of Regents

Keiki-Pua Dancil
Hawaii Science and Technology
    Institute

Reed Dasenbrock
University of Hawaii at Mānoa

David Dawson
The National Academies

Sandra Dawson
TMT Observatory Corporation

Ramon de la Pena
University of Hawaii Board of
    Regents

---

* Speakers in italics.

Greg Dickhens
Kyo-ya Hotels & Resorts

Jim Dire
Kauai Community College

Nancy Downes
Referentia Systems Inc.

Douglas Dykstra
Windward Community College

Cheryl Ernst
University of Hawaii

Thomas Ernst
Queen's Medical Center

Ardis Eschenberg
Windward Community College

Thomas Fargo
APEC Host Committee

Jane Ferreira
New Horizons CLC of Hawaii

David Ferrell

Jay Fidell
ThinkTech

Judith Flanders
OLA Hawaii 2020

Andrea Fleig
Queen's Medical Center

Craig Floro

_____

* Speakers in italics.

Guy Fo
Honolulu Community College

Yaa-Yin Fong

Karl Fooks
Hawaii Strategic Development
    Corporation

Rockne Freitas
University of Hawaii

Lynn Fujioka
IsisHawaii

Carol Fukunaga
Hawaii State Senate

Kay Fukunaga
Ulupono Initiative

Paul Fuligni
U.S. Pacific Fleet

Richard Fulton
Windward Community College

John Furstenwerth
Hawaii SBDC Network

Jim Gaines
University of Hawaii

Michelle Garcia

David Garmire
UHM College of Engineering

Gerald Garrison
Green Applied Sciences

Chuck Gee
University of Hawaii Board of Regents

Olga Geling
University of Hawaii

Daniel S. Goldin
Intellisis Corporation and
9th NASA Administrator (Ret.)

Mary Good
University of Arkansas at Little Rock

Kealii Gora
University of Hawaii at Mānoa

Peter Gorham
University of Hawaii at Mānoa

M.R.C. Greenwood
University of Hawaii

Kristina Guo
University of Hawaii at West Oahu

Donna Gutierrez
University of Hawaii Foundation

Richard Ha
HCEI

James Haley
OLA Hawaii 2020

Colleen Hanabusa
U.S. House of Representatives

Martha Hanson
University of Hawaii Foundation

Daris Hao
U.S. General Services Administration

Marguerite Harden
Kupu/R.I.S.E.

James Hardway
Workforce Development Council

William Harris
Science Foundation Arizona

Alan Hayashi
BAE Systems/MAC

Ronald Hayashi
Honolulu Japanese Chamber of
    Commerce

Jerris Hedges
John A. Burns School of Medicine
    University of Hawaii at Mānoa

Laurien Helfrich-Nuss
East West Center Fellow/OLA Hawaii
    2020

Jeannette Hereniko
OLA Hawaii 2020

Francisco Hernandez
University of Hawaii

Brandon Marc Higa
University of Hawaii

Milton Higa
Kapi'olani Community College

Virginia Hinshaw
University of Hawaii at Mānoa

Lori Hiramatsu
Hawaii Economic Development
    Corporation

Mazie Hirono
U.S. House of Representatives

Dennis Hirota
University of Hawaii Board of Regents

Peter Ho
Bank of Hawaii
and APEC 2011 Hawaii Host
   Committee

Ian Chan Hodges
American Ingenuity Alliance

Lui Hokoana

James Holm-Kennedy
University of Hawaii at Mānoa

John Holzman
University of Hawaii Board of Regents

Jacqui Hoover
Hawaii Island Economic Board

Jessica Horiuchi
Alaka'ina Foundation

Sandy Hoshino
Leeward Community College

Bernadette Howard
Windward Community College

Ching Yuan Hu
University of Hawaii at Mānoa

Keslie Hui

Lisa Hunter
Institute for Astronomy, UH

Les Ihara, Jr.
Hawaii State Senate

Daniel K. Inouye
United States Senate

Wayne Inouye
HTDC Manufacturing Extension
   Partnership

Cheryl Sato Ishii
University of Hawaii College of
   Engineering

Dan Ishii
University of Hawaii

Gayle Ishii
UHCC

Joanne Itano
University of Hawaii System

Stephen Itoga

Cameron Jaeb

Vidushi Jetley
University of Hawaii Foundation

Jeffrie Jones
University of Hawaii Foundation

Barry Johnson
Economic Development Administration
U.S. Department of Commerce

Rex Johnson
PICHTR

Linda Johnsrud

* Speakers in italics.

Greg Judd
UHM Shidler College of Business

Elmer Ka'ai
University of Hawaii at Mānoa

Philip Kahue
Ke'aki Technologies, LLC

Gary Kai
Hawaii Business Roundtable

Noelani Kalipi
TiLeaf Group

James Karins
Pukoa Scientific

Wayne Karo
Pipeline Micro

Howard Karr
University of Hawaii Board of Regents

Laurie Kawamura

Lillian Kawasaki

Abidin Kaya
Amel Technologies

Maurice Kaya
Hawaii Renewable Energy
*   Development Venture (HREDV)*

Kenneth Kelly
National Renewable Energy Laboratory

Kevin Kelly
University of Hawaii

Mark Kerber
Marine Corps Forces Pacific Science
    & Technology

Roger Kilmer
Manufacturing Extension Partnership
*   Program*
National Institute of Standards and
*   Technology*

Gregory Kim
Virtual Law Partners LLP

Guy Kimura
Hawaii Community College

Vincent Kimura

Luke Kirch
BAE Systems SSL

Ian Kitajima
Ocean it

Lianne Kitajima
TeraSys Technologies

Collin Kobayashi
3D Innovations

Mannette Kokobun

Katharine Ku
Office of Technology Licensing
Stanford University

Anthony Kuh
University of Hawaii

Jeff Kuhn
Institute for Astronomy, UH

---

* Speakers in italics.

Sandi Kwee
Queen's Medical Center

Erika Lacro
Honolulu Community College

Aaron Landry

David Lassner
University of Hawaii

Russell Lau

Anne Leake
UHM Department of Nursing

Chris Lee
Academy for Creative Media

Hye-ryeon Lee
University of Hawaii at Mānoa

Karen Lee
University of Hawaii System

Krystal Lee
Pacific Asian Center for
    Entrepreneurship

James Lee
University of Hawaii Board of Regents

Jerry S. H. Lee
Center for Strategic Scientific Initiatives
National Cancer Institute
National Institutes of Health

Mona Lee
Kapi'olani Community College

David Leonard
Solazyme, Inc.

Joyce Leung
PAAC Board Director

Ginger Lew
White House National Economic
    Council

Leslie Lewis
University of Hawaii Foundation

Richard Lim
State of Hawaii DBEDT

Dawn Lippert
PICHTR

Nadine Little
University of Hawaii Foundation

David Lonborg
University of Hawaii System

James Mackay

Justin MacNaughton

Michael Markrich
MicroPlanet

Roald Marth

Eric Martinson
University of Hawaii Board of Regents

Harold Masumoto
PICHTR

Jeff Matsu
State of Hawaii, Department of Labor

Keith Matsumoto
PICHTR

Mike May
T. Michael May Advisor Services

Kekoa McCullen

Mark McGuffie
Enterprise Honolulu

John McKee
UHMC VCAA

Kathy McKenzie
Hawaii State Energy Office

Robert McLaren
Institute for Astronomy
University of Hawaii

Mariko Miho
University of Hawaii Foundation

Jeff Mikulina
Blue Planet Foundation

Timothy Ming
Hawaii State Energy Office

Leigh-Ann Miyasato
Mānoa Venture Partners

Kevin Miyashiro
TeraSys Technologies

Daniel Momohara
Pacific Missile Range Facility

Keala Monaco
University of Hawaii System

Tyler Mongan

Lucille Moore
Tropical Telecom Corporation

Tom Moore
Hawaii Pacific Export & Chamber of
    Commerce

Lauren Moriarty
Ambassador

Melvin Morito
Wolf Creek Fabrication Services Inc.

John Morton
University of Hawaii

Scott Murakami
Pacific Center for Advanced
    Technology Training

Suzanne Murphy
University of Hawaii

Nick Murray

Craig Nakanishi
High Technology Development
    Corporation

Kabi Neupane
Leeward Community College

Susan Nishida
University of Hawaii at West Oahu

Earl Nishiguchi
Kauai Community College

Clarence Nishihara
Hawaii State Senate

---

* Speakers in italics.

James Nishimoto
University of Hawaii

Ernest Nishizaki
Kyo-ya Company, LLC

Suzanne Chun Oakland
Hawaii State Senate

Daniel O'Connell
HNU Energy

Aaron Ohta
University of Hawaii at Mānoa

Vice Admiral Daniel Oliver, USN (Ret.)
Naval Postgraduate School
Monterrey, California

Michael O'Malley
HiBEAM

Richard Ordonez
UHM College of Engineering

Charles Ota
The Chamber of Commerce Hawaii

Laurice Otsuka
American Savings Bank

Neil Oyama
PBRRTC

Constantinos Papacostas
University of Hawaii

Ann Park
UH OTTED

Patrick Park

Reinhold Penner
Queen's Medical Center

Bradley Perkins
Tissue Genesis, Inc.

Frederick Perlak
Monsanto Company

Luis Proenza
The University of Akron

Peter Quigley
UHCC

Race Randle

Assumpta Rapoza
HTDC

Teena Rasmussen
University of Hawaii Board of Regents

Michael Reiley
HNU Energy

Seth Reiss
Seth M. Reiss, AAL, ALLLC

Leon Richards
Kapi'olani Community College

Brian Richardson
Windward Community College

David Ringuette

Jay-R Rivera
Honolulu Community College

Kevin Roberts
Castle Medical Center

---

* Speakers in italics.

Trent Robertson
UHM College of Engineering

Suzette Robinson
University of Hawaii Community
    Colleges

Mark Rognstad
University of Hawaii at Mānoa

Richard Rosenblum
Hawaiian Electric Company

Nora Ruebrook
Omega

Robert Saarnio
University of Hawaii Foundation

Chhany Sak-Humphry
University of Hawaii at Mānoa

Marcia Sakai

Clyde Sakamoto
University of Hawaii Maui College

Christine Sakuda

Jane Sawyer

Brian Schatz
Lieutenant Governor of the State of
    Hawaii

Alan Schlissel
Schlissel & Associates

Margot Schrire
University of Hawaii Foundation

Kamil Schuetz

Kelly Scott
University of Hawaii Foundation

Janayhe Self

Tina Shelton
University of Hawaii System EAUR

Wayne Shiroma
UHM College of Engineering

Sujai Shivakumar
The National Academies

Lee Sichter
Belt Collins Hawaii

Joelle Simonpetri
U.S. Pacific Command

Jeanne Skog
Maui Economic Development Board

Michael Sorenson

Bill Spencer
Hawaii Oceanic Technology, Inc.

Steven Stanley
University of Hawaii

Dan Starko

Chuck Sted
Hawaii Pacific Health/HI Business
    Roundtable

Juergen Steinmetz
Hawaii Tourism Association

---

* Speakers in italics.

Amanda Stowell
University of Hawaii Foundation

Christopher Strahle
UHM School of Architecture

Donald Straney
University of Hawaii at Hilo

Tak Sugimura
University of Hawaii at Mānoa HOSC

Jan Sullivan
Oceanit

Omar Sultan

Tarik Sultan

Gerald Sumida
Carlsmith Ball LLP

David Sungarian
Sungarian Consulting

Warner Kimo Sutton
Diamond Head Renewable Resources

Matt Sweeney

Vassilis Syrmos
University of Hawaii at Mānoa

Alvin Tagomori
University of Hawaii Maui College

Joanne Taira
University of Hawaii System

Gregg Takayama
University of Hawaii at Mānoa

Alan Takemoto
Monsanto Company

Mari Takemoto-Chock

Norman Takeya
Honolulu Community College

Barbara Tanabe
Hoakea Communications

Ramsay Taum
OLA Hawaii 2020

Brian Taylor
School of Ocean and Earth Science
and Technology
University of Hawaii at Mānoa

Catherine Ann Taylor
Strategic Allies LLC

Lee Taylor
UH OTTED

Tyrone Taylor
Capitol Advisors on Technology

Yukiyo Tetsumura

Pete Thompson
Morgan Stanley Smith & Barney

Howard Todo
University of Hawaii

Jill Tokuda
Hawaii State Senate

Jim Tollefson
The Chamber of Commerce of Hawaii

---

* Speakers in italics.

Sunshine Topping
State of Hawaii, Department of
    Human Resources

Shanah Trevenna

Rose Tseng
University of Hawaii at Hilo

Noble Turner

Art Ushijima
Queen's Health Systems/The Queen's
    Medical Center

Allen Uyeda
First Insurance Company

Mike Von Fahnestock
U.S. Pacific Command

Starnes Walker
University of Hawaii

Mary Walshok
University of California at San Diego

Barry Weinman
Allegis Capital LLC

Charles Wessner
The National Academies

Hank Wuh
Skai Ventures
and Cellular Bioengineering, Inc.

Sylvia Yuen
College of Tropical Agriculture and
    Human Resources
University of Hawaii at Mānoa

---

* Speakers in italics.

Appendix D

Bibliography

Acs, Z., and D. Audretsch. 1990. *Innovation and Small Firms*. Cambridge, MA: The MIT Press.

Alic, J. A., L. M. Branscomb, H. Brooks, A. B. Carter, and G. L. Epstein. 1992. *Beyond Spin-off: Military and Commercial Technologies in a Changing World*. Boston: Harvard Business School Press.

Amsden, A. H. 2001. *The Rise of "the Rest": Challenges to the West from Late-industrializing Economies*. Oxford: Oxford University Press.

Asheim, B., A. Isaksen, C. Nauwelaers, and F. Todtling, eds. 2003. *Regional Innovation Policy for Small-Medium Enterprises*. Cheltenham, UK, and Northampton, MA: Edward Elgar.

Athreye, S. 2000. "Technology Policy and Innovation: The Role of Competition Between Firms." In P. Conceicao, S. Shariq, and M. Heitor, eds. *Science, Technology, and Innovation Policy: Opportunities and Challenges for the Knowledge Economy*. Westport, CT, and London: Quorum Books.

Atkinson, R., and S. Andes. 2010. *The 2010 State New Economy Index: Benchmarking Economic Transformation in the States*. Kauffman Foundation and ITIF. November.

Audretsch, D., ed. 1998. *Industrial Policy and Competitive Advantage, Volumes 1 and 2*. Cheltenham, UK: Edward Elgar.

Audretsch, D. 2006. *The Entrepreneurial Society*, Oxford: Oxford University Press.

Audretsch, D., B. Bozeman, K. L. Combs, M. Feldman, A. Link, D. Siegel, P. Stephan, G. Tassey, and C. Wessner. 2002. "The economics of science and technology." *Journal of Technology Transfer* 27:155–203.

Audretsch, D., H. Grimm, and C. W. Wessner. 2005. *Local Heroes in the Global Village: Globalization and the New Entrepreneurship Policies*. New York: Springer.

Augustine, C., et al. 2009. *Redefining What's Possible for Clean Energy by 2020*. Full Report. Gigaton Throwdown. June.

Bajaj, V. 2009. "India to spend $900 million on solar." *The New York Times* November 20.

Baldwin, J. R., and P. Hanel. 2003. *Innovation and Knowledge Creation in an Open Economy: Canadian Industry and International Implications*. Cambridge: Cambridge University Press.

Balzat, M., and A. Pyka. 2006. "Mapping national innovation systems in the OECD area." *International Journal of Technology and Globalisation* 2(1-2):158-176.

176

Bezdek, R. H., and F. T. Sparrow. 1981. "Solar subsidies and economic efficiency." *Energy Policy* 9(4):289-300.

Biegelbauer, P. S., and S. Borras, eds. 2003. *Innovation Policies in Europe and the U.S.: The New Agenda*. Aldershot, UK: Ashgate.

Birch, D. 1981. "Who creates jobs?" *The Public Interest* 65:3-14.

Blomström, M., A. Kokko, and F. Sjöholm. 2002. "Growth & Innovation Policies for a Knowledge Economy: Experiences from Finland, Sweden, & Singapore." EIJS Working Paper. Series No. 156.

Bloomberg News. 2006. "The next green revolution." August 21.

Bolinger, M., R. Wiser, and E. Ing. 2006. "Exploring the Economic Value of EPAct 2005's PV Tax Credits." Lawrence Berkeley National Laboratory.

Borenstein, S. 2008. *The Market Value and Cost of Solar Photovoltaic Electricity Production*. Berkeley, CA: Center for the Study of Energy Markets.

Borras, S. 2003. *The Innovation Policy of the European Union: From Government to Governance*. Cheltenham, UK: Edward Elgar.

Borrus, M., and J. Stowsky. 2000. "Technology policy and economic growth." In C. Edquist and M. McKelvey, eds. *Systems of Innovation: Growth, Competitiveness and Employment*, Vol. 2. Cheltenham, UK and Northampton, MA: Edward Elgar.

Bradsher, K. 2009. "China builds high wall to guard energy industry." *International Herald Tribune* July 13.

Brander, J. A., and B. J. Spencer. 1983. "International R&D rivalry and industrial strategy." *Review of Economic Studies* 50:707-722.

Brander, J. A., and B. J. Spencer. 1985. "Export strategies and international market share rivalry." *Journal of International Economics* 16:83-100.

Branigin, W. 2009. "Obama lays out clean-energy plans." *Washington Post* March 24, p. A05.

Branscomb, L., and P. Auerswald. 2002. *Between Invention and Innovation: An Analysis of Funding for Early-Stage Technology Development*. NIST GCR 02–841. Gaithersburg, MD: National Institute of Standards and Technology. November.

Braunerhjelm, Pontus and Maryann Feldman. 2006. *Cluster Genesis: Technology based Industrail Development*. Oxford: Oxford University Press.

Bush, N. 2005. "Chinese competition policy, it takes more than a law." *China Business Review* May-June.

Bush, V. 1945. *Science: The Endless Frontier*. Washington, D.C.: Government Printing Office.

Campoccia, A., L. Dusonchet, E. Telaretti, and G. Zizzo. 2009. "Comparative analysis of different supporting measures for the production of electrical energy by solar PV and Wind systems: Four representative European cases." *Solar Energy* 83(3):287-297.

Caracostas, P., and U. Muldur. 2001. "The emergence of the new European Union research and innovation policy." In P. Laredo and P. Mustar, eds. *Research and Innovation Policies in the New Global Economy: An International Comparative Analysis*. Cheltenham, UK: Edward Elgar.

Chesbrough, H. 2003. *Open Innovation: The New Imperative for Creating and Profiting from Technology*. Cambridge, MA: Harvard Business School Press.

Cimoli, M., and M. della Giusta. 2000. "The Nature of Technological Change and Its Main Implications on National and Local Systems of Innovation." IIASA Interim Report IR-98-029.

Coburn, C., and D. Berglund. 1995. *Partnerships: A Compendium of State and Federal Cooperative Programs*. Columbus, OH: Battelle Press.

Combs, K., and A. Link. 2003. "Innovation policy in search of an economic paradigm: the case of research partnerships in the United States." *Technology Analysis & Strategic Management* 15(2).

Cortright, J. 2006. *Making Sense of Clusters: Regional Competitiveness and Economic Development*. Washington, D.C.: Brookings Institution.

Cortright, J., and Mayer, H. 2002. *Signs of Life: The Growth of Biotechnology Centers in the United States*. Washington, D.C.: Brookings Institution.

Crafts, N. F. R. 1995. "The golden age of economic growth in Western Europe, 1950–1973." *Economic History Review* 3:429-447.

Dahlman, C., and J. E. Aubert. 2001. *China and the Knowledge Economy: Seizing the 21st Century*. Washington, D.C.: World Bank.

Dahlman, C., and A. Utz. 2005. *India and the Knowledge Economy: Leveraging Strengths and Opportunities*. Washington, D.C.: World Bank.

Davis, S., J. Haltiwanger, and S. Schuh. 1993. "Small Business and Job Creation: Dissecting the Myth and Reassessing the Facts." Working Paper No. 4492. Cambridge, MA: National Bureau of Economic Research.

Debackere, K., and R. Veugelers. 2005. "The role of academic technology transfer organizations in improving industry science links." *Research Policy* 34(3):321-342.

Department of Labor and Industrial Relations: Research and Statistics Office. *Hawaii's Green Workforce: A Baseline Assessment*. December 2010.

DeVol, Ross C., Kevin Klowden, Armen Bedorussian, and Benjamin Yeo. 2009. *North America's High Tech Economy: The Geography of Knowledge-Based Institututions*. June 2.

De la Mothe, J., and G. Paquet. 1998. "National Innovation Systems, 'Real Economies' and Instituted Processes." *Small Business Economics* 11:101-111.

Dobesova, K., J. Apt, and L. Lave. 2005. "Are renewable portfolio standards cost-effective emissions abatement policy?" *Environmental Science and Technology* 39:8578-8583.

Doloreux, D. 2004. "Regional innovation systems in Canada: a comparative study." *Regional Studies* 38(5):479-492.

Doris, E., J. McLaren, V. Healey, and S. Hockett. 2009. *State of the States 2009: Renewable Energy Development and the Role of Policy*. Golden, CO: National Renewable Energy Laboratory.

Durham, C. A., B. G. Colby, M. Longstreth. 1988. "The impact of state tax credits and energy prices on adoption of solar energy systems." *Land Economics* 64(4):347-355.

Eaton, J., E. Gutierrez, and S. Kortum. 1998. "European Technology Policy." NBER Working Paper 6827.

Edler, J., and S. Kuhlmann. 2005. "Towards one system? The European Research Area initiative, the integration of research systems and the changing leeway of national policies." *Technikfolgenabschätzung: Theorie und Praxis* 1(4):59-68.

Eickelpasch, A., and M. Fritsch. 2005. "Contests for cooperation: a new approach in German innovation policy." *Research Policy* 34:1269-1282.

Energy Information Administration. 2008. *Federal Financial Interventions and Subsidies in Energy Markets 2007*. Washington, D.C.: Energy Information Administration.

European Commission. 2003. "Innovation in Candidate Countries: Strengthening Industrial Performance." Brussels: European Commission. May.

Fangerberg, J. 2002. *Technology, Growth, and Competitiveness: Selected Essays*. Cheltenham, UK, and Northampton, MA: Edward Elgar.

Federal Reserve of Chicago. 2007. "Can Higher Education Foster Economic Growth?—A Conference Summary. *Chicago Fed Letter*. March.

Feldman, M., and A. Link. 2001. "Innovation policy in the knowledge-based economy." In *Economics of Science, Technology and Innovation, Vol. 23*. Boston: Kluwer Academic Press.

Feldman, M., A. Link, and D. Siegel. 2002. *The Economics of Science and Technology: An Overview of Initiatives to Foster Innovation, Entrepreneurship, and Economic Growth*. Boston: Kluwer Academic Press.

Feser, Edward. 2005. "Industry Cluster Concepts in Innovation Policy: A Comparison of U.S. and Latin American Experience." *Interdiscliplinary Studies in Economics and Management*. Volume 4. Vienna: Springer.

Flamm, K. 2003. "SEMATECH revisited: assessing consortium impacts on semiconductor industry R&D." In National Research Council. *Securing the Future: Regional and National Programs to Support the Semiconductor Industry.* C. W. Wessner, ed. Washington, D.C.: The National Academies Press.

Fonfria, A., C. Diaz de la Guardia, and I. Alvarez. 2002. "The role of technology and competitiveness policies: a technology gap approach." *Journal of Interdisciplinary Economics* 13:223-241.

Foray, D., and P. Llerena. 1996. "Information structure and coordination in technology policy: a theoretical model and two case studies." *Journal of Evolutionary Economics* 6(2):157-173.

Fishback, Bo, Christine A. Gulbranson, Robert E. Litan, Lesa Mitchell and Marisa Porzig. 2007. *Finding Business "Idols": A New Model to Accelerate Start-Ups*, Kauffman Foundation Report, 4.

Florida, Richard. *The Rise of the Creative Class.* New York: Basic Books. 2002.

Friedman, T. 2005. *The World Is Flat: A Brief History of the 21st Century.* New York: W. H. Freeman.

Fry, G. R. H. 1986. "The economics of home solar water heating and the role of solar tax credits." *Land Economics* 62(2):134-144.

Fthenakis, V., J. E. Mason, and K. Zweibel. 2009. "The technical, geographical, and economic feasibility for solar energy to supply the energy needs of the US." *Energy Policy* 37(2):387-399.

Fullilove, Mindy Thompson. 2005. *Root Shock: How Tearing Up City Neighborhoods Hurts America and What We Can Do About It.* New York: Ballantine Books.

Furman, J., M. Porter, and S. Stern. 2002. "The determinants of national innovative capacity." *Research Policy* 31:899-933.

Geiger, Roger L. and Creso M. Sá. 2009. *Tapping the Riches of Science: Universities and the Promise of Economic Growth.* Cambridge, MA: Harvard University Press.

George, G., and G. Prabhu. 2003. "Developmental financial institutions as technology policy instruments: implications for innovation and entrepreneurship in emerging economies." *Research Policy* 32(1):89-108.

Grande, E. 2001. "The erosion of state capacity and European innovation policy: a comparison of German and EU information technology policies." *Research Policy* 30(6):905-921.

Grindley, P., D. Mowery, and B. Silverman. 1994. "SEMATECH and collaborative research: lessons in the design of high technology consortia." *Journal of Policy Analysis and Management* 13(4):723-758.

Grossman, G. M., and E. Helpman. 1994. "Endogenous innovation in the theory of growth." *The Journal of Economic Perspectives* 8(1):23-44.

Guidolin, M., and C. Mortarino. 2010. "Cross-country diffusion of photovoltaic systems: modelling choices and forecasts for national adoption patterns." *Technological Forecasting and Social Change* 77(2):279-296.

Hall, B. 2002. "The assessment: technology policy." *Oxford Review of Economic Policy* 18(1):1-9.

Hall, Bronwyn. 2004. "University-Industry Research Partnerships in the United States." Kansai Symposium Report. February.

Hu, Z. 2006. "IPR Policies In China: Challenges and Directions." Presentation at *Industrial Innovation in China*. Levin Institute Conference. July 24-26.

Hughes, K. 2005. *Building the Next American Century: The Past and Future of American Economic Competitiveness.* Washington, D.C.: Woodrow Wilson Center Press.

Hughes, K. 2005. "Facing the global competitiveness challenge." *Issues in Science and Technology* XXI(4):72-78.

Jaffe, A., J. Lerner, and S. Stern, eds. 2003. *Innovation Policy and the Economy, Vol. 3.* Cambridge, MA: MIT Press.

Janssen, Marco A, Robert Holahan, Allen Lee, and Elinor Ostrom. 2010. "Lab Experiments for the Study of Social-Ecological Systems." *Science* 328(5978):613-617. April.

Jaruzelski, Barry, and Kevin Dehoff. 2008. "Beyond Borders: The Global Innovation 1000." *Strategy and Business* 53(Winter).

Jasanoff, S., ed. 1997. *Comparative Science and Technology Policy.* Elgar Reference Collection. International Library of Comparative Pubic Policy, Vol. 5. Cheltenham, UK, and Lyme, NH: Edward Elgar.

Jorgenson, D., and K. Stiroh. 2002. "Raising the speed limit: economic growth in the information age." In National Research Council. 2002. *Measuring and Sustaining the New Economy.* D. W. Jorgenson and C. W. Wessner, eds. Washington, D.C.: The National Academies Press.

Joy, W. 2000. "Why the future does not need us." *Wired* 8(April).

Kim, Yong-June. 2006. "A Korean Perspective on China's Innovation System." Presentation at *Industrial Innovation in China.* Levin Institute Conference. July 24-26.

Koschatzky, K. 2003. "The regionalization of innovation policy: new options for regional change?" In G. Fuchs and P. Shapira, eds. *Rethinking Regional Innovation: Path Dependency or Regional Breakthrough?* London: Kluwer.

Krueger, Anne O. "Globalization and International Locational Competition." Symposium in Honor of Herbert Giersch. Lecture delivered at the Keil Institute. May 11, 2006.

Kuhlmann, S., and J. Edler. 2003. "Scenarios of technology and innovation policies in Europe: investigating future governance—group of 3." *Technological Forecasting & Social Change* 70.

Lall, S. 2002. "Linking FDI and technology development for capacity building and strategic competitiveness." *Transnational Corporations* 11(3):39-88.

Lancaster, R. R., and M. J. Berndt. 1984. "Alternative energy development in the USA: the effectiveness of state government incentives." *Energy Policy* 12(2):170-179.

Laredo, P., and P. Mustar, eds. 2001. *Research and Innovation Policies in the New Global Economy: An International Perspective.* Cheltenham, UK: Edward Elgar.

Lee, Y. S. 2000. "The Sustainability of University-Industry Research Collaboration. *The Journal of Technology Transfer* 25(2).

Lerner, J. 1999. "Public venture capital." In National Research Council. *The Small Business Innovation Program: Challenges and Opportunities.* C. W. Wessner, ed. Washington, D.C.: National Academy Press.

Lewis, J. 2005. *Waiting for Sputnik: Basic Research and Strategic Competition.* Washington, D.C.: Center for Strategic and International Studies.

Lin, O. 1998. "Science and technology policy and its influence on the economic development of Taiwan." In H. S. Rowen, ed. *Behind East Asian Growth: The Political and Social Foundations of Prosperity.* London and New York: Routledge.

Link, Albert N. 1995. *A Generosity of Spirit: The Early History of the Research Triangle Park.* Research Triangle Park: The Research Triangle Foundation of North Carolina.

Litan, Robert E., Lesa Mitchell and E. J. Reedy 2007. "The University as Innovator: Bumps in the Road." *Issues in Science and Technology* Summer, 57-66.

Lucas, Robert. "On the mechanics of economic development." *Journal of Military Economics* 22:38-39.

Luger, M. 2001. "Introduction: information technology and regional economic development." *Journal of Comparative Policy Analysis: Research & Practice.*

Luger, M., and H. A. Goldstein. 1991. *Technology in the Garden.* Chapel Hill: University of North Carolina Press.

Luger, M., and H. A. Goldstein. 2006. *Research Parks Redux: The Changing Landscape of the Garden.* Washington, D.C.: U.S. Department of Commerce.

Luther, J. 2008. "Renewable Energy Development in Germany." Presentation and the NRC Christine Mirzayan Fellows Seminar. March 5, 2008. Washington, D.C.

Maddison, A., and D. Johnston. 2001. *The World Economy: A Millennial Perspective.* Paris: Organisation for Economic Co-operation and Development.

Mani, S. 2004. "Government, innovation and technology policy: an international comparative analysis." *International Journal of Technology and Globalization* 1(1).

McKibben, W. 2003. *Enough: Staying Human in an Engineered Age.* New York: Henry Holt & Co.

Mendonca, M. 2007. *Feed-in Tariffs: Accelerating the Development of Renewable Energy*. London: Earthscan.

Meyer-Krahmer, F. 2001. Industrial innovation and sustainability—conflicts and coherence." In D. Archibugi and B. Lundvall, eds. *The Globalizing Learning Economy*. New York: Oxford University Press.

Meyer-Krahmer, F. 2001. "The German innovation system." In P. Larédo and P. Mustar, eds. *Research and Innovation Policies in the New Global Economy: An International Comparative Analysis*. Cheltenham, UK: Edward Elgar.

Mills, K. G., E. B. Reynolds, and A. Reamer. 2008. *Clusters and Competitiveness: A New Federal Role for Stimulating Regional Economies*. Washington, D.C.: Brookings.

Moore, G. 2003. "The SEMATECH contribution." In National Research Council. *Securing the Future: Regional and National Programs to Support the Semiconductor Industry*. C. W. Wessner, ed. Washington, D.C.: The National Academies Press.

Moselle, B., J. Padilla, and R. Schmalensee. 2010. *Harnessing Renewable Energy in Electric Power Systems: Theory, Practice Policy*. Washington, D.C.: RFF Press.

Mufson, S. 2009. "Asian nations could outpace U.S. in developing clean energy." *Washington Post* July 16.

Murphy, L. M., and P. L. Edwards. 2003. *Bridging the Valley of Death: Transitioning from Public to Private Sector Financing*. Golden, CO: National Renewable Energy Laboratory. May.

Mustar, P., and P. Laredo. 2002. "Innovation and research policy in France (1980-2000) or the disappearance of the Colbertist state." *Research Policy* 31:55-72.

National Academy of Engineering. 2004. *The Engineer of 2020: Visions of Engineering in the New Century*. Washington, D.C.: The National Academies Press.

National Academy of Engineering. 2008. *Grand Challenges for Engineering*. Washington, D.C.: The National Academies Press.

National Academy of Sciences. 2010. *Electricity from Renewable Sources: Status, Prospects, and Impediments*. Washington, D.C.: The National Academies Press.

National Academy of Sciences, National Academy of Engineering, and Institute of Medicine. 2007. *Rising Above the Gathering Storm: Energizing and Employing America for a Brighter Economic Future*. Washington, D.C.: The National Academies Press.

National Academy of Sciences, National Academy of Engineering, and National Research Council. 2009. *America's Energy Future: Technology and Transformation*. Washington, D.C.: The National Academies Press.

National Academy of Sciences, National Academy of Engineering, and National Research Council. 2009. *Real Prospects for Energy Efficiency in the United States*. Washington, D.C.: The National Academies Press.

National Governors' Association. 2007. *Innovation America*. Washington, D.C.: National Governors' Association.

National Research Council. 1996. *Conflict and Cooperation in National Competition for High-Technology Industry*. Washington, D.C.: National Academy Press.

National Research Council. 1999. *The Advanced Technology Program: Challenges and Opportunities*. C. W. Wessner, ed. Washington, D.C.: National Academy Press.

National Research Council. 1999. *Funding a Revolution: Government Support for Computing Research*. Washington, D.C.: National Academy Press.

National Research Council. 1999. *Industry-Laboratory Partnerships: A Review of the Sandia Science and Technology Park Initiative*. C. W. Wessner, ed. Washington, D.C.: National Academy Press.

National Research Council. 1999. *New Vistas in Transatlantic Science and Technology Cooperation*. C. W. Wessner, ed. Washington, D.C.: National Academy Press.

National Research Council. 1999. *The Small Business Innovation Research Program: Challenges and Opportunities*. C. W. Wessner, ed. Washington, D.C.: National Academy Press.

National Research Council. 1999. *U.S. Industry in 2000: Studies in Competitive Performance.* D, C. Mowery, ed. Washington, D.C.: National Academy Press.

National Research Council. 2000. *The Small Business Innovation Research Program: A Review of the Department of Defense Fast Track Initiative.* C. W. Wessner, ed. Washington, D.C.: National Academy Press.

National Research Council. 2001. *A Review of the New Initiatives at the NASA Ames Research Center.* C. W. Wessner, ed. Washington, D.C.: National Academy Press.

National Research Council. 2001. *Building a Workforce for the Information Economy.* Washington, D.C.: National Academy Press.

National Research Council. 2001. *Capitalizing on New Needs and New Opportunities: Government-Industry Partnerships in Biotechnology and Information Technologies.* C. W. Wessner, ed. Washington, D.C.: National Academy Press.

National Research Council. 2001. *The Advanced Technology Program: Assessing Outcomes.* C. W. Wessner, ed. Washington, D.C.: National Academy Press.

National Research Council. 2001. *Trends in Federal Support of Research and Graduate Education.* S. A. Merrill, ed. Washington, D.C.: National Academy Press.

National Research Council. 2003. *Partnerships for Solid-State Lighting.* C. W. Wessner, ed. Washington, D.C.: The National Academies Press.

National Research Council. 2003. *Government-Industry Partnerships for the Development of New Technologies: Summary Report.* C. W. Wessner, ed. Washington, D.C.: The National Academies Press.

National Research Council. 2003. *Securing the Future: Regional and National Programs to Support the Semiconductor Industry.* C. W. Wessner, ed. Washington, D.C.: The National Academies Press.

National Research Council. 2003. *Understanding Climate Change Feedbacks*, Washington, D.C.: The National Academies Press.

National Research Council. 2004. *The Small Business Innovation Research Program: Program Diversity and Assessment Challenges.* C. W. Wessner, ed. Washington, D.C.: The National Academies Press.

National Research Council. 2005. *Getting Up to Speed: The Future of Superconducting.* S. L. Graham, M. Snir, and C. A. Patterson, eds. Washington, D.C.: The National Academies Press.

National Research Council. 2005. *Policy Implications of International Graduate Students and Post-doctoral Scholars in the United States.* Washington, D.C.: The National Academies Press.

National Research Council. 2007. *Enhancing Productivity Growth in the Information Age: Measuring and Sustaining the New Economy.* D. W. Jorgenson and C. W. Wessner, eds. Washington, D.C.: The National Academies Press.

National Research Council. 2007. *Innovation Policies for the 21st Century.* C. W. Wessner, ed. Washington, D.C.: The National Academies Press.

National Research Council. 2007. *India's Changing Innovation System: Achievements, Challenges, and Opportunities for Cooperation.* C. W. Wessner and S. J. Shivakumar, eds. Washington, D.C.: The National Academies Press.

National Research Council. 2007. *SBIR and the Phase III Challenge of Commercialization.* C. W. Wessner, ed. Washington, D.C.: The National Academies Press.

National Research Council. 2008. *An Assessment of the Small Business Innovation Research Program.* C. W. Wessner, ed. Washington, D.C.: The National Academies Press.

National Research Council. 2008. *An Assessment of the Small Business Innovation Research Program at the Department of Energy.* C. W. Wessner, ed. Washington, D.C.: The National Academies Press.

National Research Council. 2008. *An Assessment of the Small Business Innovation Research Program at the National Science Foundation.* C. W. Wessner, ed. Washington, D.C.: The National Academies Press.

National Research Council. 2008. *Innovative Flanders: Innovation Policies for the 21st Century*. C. W. Wessner, ed. Washington, D.C.: The National Academies Press.

National Research Council. 2008. *Innovation in Global Industries: U.S. Firms Competing in a New World*. J. Macher and D. Mowery, eds. Washington D.C.: The National Academies Press.

National Research Council. 2008. *The National Academies Summit on America's Energy Future: Summary of a Meeting*. Washington, D.C.: The National Academies Press.

National Research Council. 2009. *21st Century Innovation Systems for Japan and the United States: Lessons from a Decade of Change*. S. Nagaoka, M. Kondo, K. Flamm, and C. Wessner, eds. Washington, D.C.: The National Academies Press.

National Research Council. 2009. *An Assessment of the Small Business Innovation Research Program at the Department of Defense*. C. W. Wessner, ed. Washington, D.C.: The National Academies Press.

National Research Council. 2009. *An Assessment of the Small Business Innovation Research Program at the National Aeronautics and Space Administration*. C. W. Wessner, ed. Washington, D.C.: The National Academies Press.

National Research Council. 2009. *An Assessment of the Small Business Innovation Research Program at the National Institutes of Health*. C. W. Wessner, ed. Washington, D.C.: The National Academies Press.

National Research Council. 2009. *Hidden Costs of Energy: Unpriced Consequences of Energy Production and Use*. Washington, D.C.: The National Academies Press.

National Research Council. 2009. *Revisiting the Department of Defense SBIR Fast Track Initiative*. C. W. Wessner, ed. Washington, D.C.: The National Academies Press.

National Research Council. 2009. *Understanding Research, Science and Technology Parks: Global Best Practices*. C. W. Wessner, ed. Washington, D.C.: The National Academies Press.

National Research Council. 2009. *Venture Funding and the NIH SBIR Program*. C. W. Wessner, ed. Washington, D.C.: The National Academies Press.

National Research Council. 2010. *Managing University Intellectual Property in the Public Interest*, Stephen Merrill and Anne-Marie Mazza, eds. Washington, D.C.: The National Academies Press.

National Research Council. 2011. *Building the 21st Century: U.S.-China Cooperation on Science, Technology, and Innovation*. C. W. Wessner, rapporteur. Washington, D.C.: The National Academies Press.

National Research Council. 2011. *Growing Innovation Clusters for American Prosperity: Summary of a Symposium*, Charles W. Wessner, ed., Washington, D.C.: The National Academies Press.

National Research Council. 2011. *The Future of Photovoltaics Manufacturing in the United States*. C. W. Wessner, rapporteur. Washington, D.C.: The National Academies Press.

Needham, J. 1954-1986. *Science and Civilization in China* (five volumes). Cambridge: Cambridge University Press.

Nelson, R., and K. Nelson. 2002. "Technology, institutions, and innovation systems." *Research Policy* 31:265-272.

Nelson, R., and N. Rosenberg. 1993. "Technical innovation and national systems." In R. R. Nelson, ed. *National Innovation Systems: A Comparative Analysis*. Oxford: Oxford University Press.

O'Hara, Margaret Pugh. 2005. *Cities of Knowledge: Cold War Science and the Search for the Next Silicon Valley*. Princeton: Princeton University Press.

Organisation for Economic Co-operation and Development. 2009. *Main Science and Technology Indicators*. Paris: Organisation for Economic Co-operation and Development.

Orszag, P., and T. Kane. 2003. "Funding Restrictions at Public Universities: Effects and Policy Implications." *Brookings Institution Working Paper*. September.

Oughton, C., M. Landabaso, and K. Morgan. 2002. "The regional innovation paradox: innovation policy and industrial policy." *Journal of Technology Transfer* 27(1).

Palmintera, Diane, 2005. "Accelerating Economic Development through University Technology Transfer." Diane Palmintera Innovation Associates.

Pavitt, K. 1998. "The Social Shaping of the National Science Base." *Research Policy* 27:793-805.

Porter, Michael E. 1998. "Clusters and the new economics of competition" *Harvard Business Review* 76(6):77-90.

Posen, A. 2001. "Japan." In B. Steil, D. G. Victor, and R. R. Nelson, eds. *Technological Innovation and Economic Performance*. Princeton: Princeton University Press.

President's Council of Advisors on Science and Technology. 2004. "Sustaining the Nation's Innovation System: Report on Information Technology Manufacturing and Competitiveness." Washington, D.C.: Executive Office of the President. January.

PricewaterhouseCoopers. 2006. "China's Impact on the Semiconductor Industry: 2005 Update." PricewaterhouseCoopers.

Raduchel, William. 2006. "The end of stovepiping." In National Research Council. *The Telecommunications Challenge: Changing Technologies and Evolving Policies*, C. W. Wessner, ed., Washington, D.C.: The National Academies Press.

Ragwitz, M., and C. Huber. 2005. "Feed-in systems in Germany and Spain: a comparison." Fraunhofer Institut für Systemtechnik und Innovationsforschung.

Reid, T. R. 2004. *The United States of Europe: The New Superpower and the End of American Supremacy*. New York: Penguin Press.

Renewable Energy Policy Network for the 21st Century. 2009. *Renewables Global Status Report 2009*. Paris: REN21.

Rickerson, W., and R. Grace. 2007. "The Debate Over Fixed Price Incentives for Renewable Electricity in Europe and the United States: Fallout and Future Directions." White Paper prepared for the Heinrich Böll Foundation. Washington, D.C.

Romer, P. M. 1990. "Endogenous technological change." *Journal of Political Economy* October.

Rosenberg, N. and R. R. Nelson. 1994. "American universities and technical advance in industry." *Research Policy* 23:323-248.

Ruttan, V. 2002. *Technology, Growth and Development: An Induced Innovation Perspective*. Oxford: Oxford University Press.

Rutten, R., and F. Boekema. 2005. "Innovation, policy and economic growth: theory and cases." *European Planning Studies* 13(8).

Sallet, Jonathan, Ed Paisley and Justin R. Masterman. 2009. "The Geography of Innovation, the Federal Government and the Growth of Regional Innovation clusters." *Science Progress.* September.

Sarzynski, A. 2010. "The Impact of Solar Incentive Programs in Ten States." George Washington Institute of Public Policy Technical Report. Revised March 2010.

Saxenian, AnnaLee. 1994. *Regional Advantage: Culture and Competition in Silicon Valley and Route 128.* Cambridge, MA: Harvard University Press.

Scott, Allen J. 2004. *On Hollywood: The Place, the Industry.* Princeton: Princeton University Press.

SERI (Solar Energy Industries Association). 2009. *U.S. Solar in Review 2008.* Washington, D.C.: Solar Energy Industries Association.

Shang, Y. 2006. "Innovation: New National Strategy of China." Presentation at Industrial Innovation in China. Levin Institute Conference. July 24-26.

Sheehan, J., and A. Wyckoff. 2003. "Targeting R&D: Economic and Policy Implications of Increasing R&D Spending." DSTI/DOC(2003)8. Paris: Organisation for Economic Co-operation and Development.

Sherwood, L. 2008. *U.S. Solar Market Trends 2007.* Latham, NY: Interstate Renewable Energy Council.

Smits, R., and S. Kuhlmann. 2004. "The rise of systemic instruments in innovation policy." *International Journal of Foresight and Innovation Policy* 1(1/2).

Speck, S. 2008. "The design of carbon and broad-based energy taxes in European countries." *Vermont Journal of Environmental Law* 10.

Spencer, W., and T. E. Seidel. 2004. "International technology roadmaps: The U.S. semiconductor experience." In National Research Council. *Productivity and Cyclicality in Semiconductors: Trends, Implications, and Questions.* D. W. Jorgenson and Charles W. Wessner, eds. Washington, D.C.: The National Academies Press.

Stanford University. 1999. *Inventions, Patents and Licensing: Research Policy Handbook.* Document 5.1. July 15.

Stokes, D. E. 1997. *Pasteur's Quadrant: Basic Science and Technological Innovation,* Washington, D.C.: Brookings Institution.

Sturgeon, T. J. 2000. "How Silicon Valley Came to Be." In M. Kenney, ed., *Understanding Silicon Valley: The Anatomy of an Entrepreneurial Region* (pp. 15-47). Stanford, CA: Stanford University Press.

Taleb, Nassim Nicholas. 2007. *The Black Swan: The Impact of the Highly Improbable.* New York: Random House.

Tan, Justin. 2006. "Growth of industry clusters and innovation: lessons from Beijing Zhongguancun Science Park." *Journal of Business Venturing* 21(6):827-850. November.

Tassey, G. 2004. "Policy issues for R&D investment in a knowledge-based economy." *Journal of Technology Transfer* 29:153-185.

Taylor, M. 2008. "Beyond technology-push and demand-pull: lessons from California's solar policy." *Energy Economics* 30(6):2829-2854.

Teubal, M. 2002. "What is the systems perspective to innovation and technology policy and how can we apply it to developing and newly industrialized economies?" *Journal of Evolutionary Economics* 12(1-2).

Tödtling, Franz, and Michaela Trippl. 2005 "One size fits all? Towards a differentiated regional innovation policy approach." *Research Policy* 34.

Tol, R. S. J. 2008. "The social cost of carbon: trends, outliers, and catastrophes." *Economics—the Open-Access, Open-Assessment E-Journal* 2(25):1-24.

Tzang, Cheng-Hua. 2010. "Managing innovation for economic development in greater China: The origins of Hsinchu and Zhongguancun." *Technology in Society* 32(2):110-121. May.

U.S. Department of Energy. 2006. Press Release. "Department Requests $4.1 Billion Investment as Part of the American Competitiveness Initiative: Funding to support Basic Scientific Research." February 2.

U.S. General Accounting Office. 2002. *Export Controls: Rapid Advances in China's Semiconductor Industry Underscore need for Fundamental U.S. Policy Review.* GAO-020620. Washington, D.C.: U.S. General Accounting Office. April.

Van Looy, B., M. Ranga, J. Callaert, K. Debackereand, and E. Zimmermann. 2004. "Combining entrepreneurial and scientific performance in academia: towards a compounded and reciprocal Matthew-effect?" *Research Policy* 33(3):425-441.

Veugelers, R., J. Larosse, M. Cincera, D. Carchon, and R. Kalenga-Mpala. 2004. "R&D activities of the business sector in Flanders: results of the R&D surveys in the context of the 3% target." Brussels: IWT-Studies.

Wang, C. 2005. "IPR sails against current stream." *Caijing* October 17.

Wang, Q. 2010. "Effective policies for renewable energy—the example of China's wind power— lessons for China's photovoltaic power." *Renewable and Sustainable Energy Reviews* 14(2):702-712.

Wessner, C. W. 2005. "Entrepreneurship and the innovation ecosystem." In D. B. Audretsch, H. Grimm, and C. W. Wessner, eds. *Local Heroes in the Global Village: Globalization and the New Entrepreneurship Policies.* New York: Springer.

Wessner, C. W. Forthcoming. *Growing Innovation Clusters for American Prosperity.* Washington, D.C.: The National Academies Press.

Wessner, C. W. 2005. *Partnering Against Terrorism.* Washington, D.C.: The National Academies Press.

Wiser, R., G. Barbose, C. Peterman, and N. Darghouth. 2009. *Tracking the Sun II: The Installed Cost of Photovoltaics in the U.S. From 1998-2008*. Berkeley, CA: Lawrence Berkeley National Laboratory.

Witt, C. E., R. L. Mitchell, and G. D. Mooney. 1993. "Overview of the Photovoltaic Manufacturing Technology (PVMaT) Project." Paper presented at the 1993 National Heath Transfer Conference. August 8-11, 1993. Atlanta, Georgia.

Yu, Junbo, and Randall Jackson. 2011. "Regional Innovation Clusters: A Critical Review." *Growth and Change* 42(2).

Zeigler, N. 1997. *Governing Ideas: Strategies for Innovation in France and Germany*. Ithaca, NY, and London: Cornell University Press.

Zweibel, K. 2010. "Should solar photovoltaics be deployed sooner because of long operating life at low, predictable cost?" *Energy Policy* 38(11):7519-7530.